Surface Modification of Metals and Alloys

Surface Modification of Metals and Alloys

Editors

Oriol Rius-Ayra
Nuria Llorca-Isern

MDPI • Basel • Beijing • Wuhan • Barcelona • Belgrade • Manchester • Tokyo • Cluj • Tianjin

Editors
Oriol Rius-Ayra
Universitat de Barcelona
Spain

Nuria Llorca-Isern
Universitat de Barcelona
Spain

Editorial Office
MDPI
St. Alban-Anlage 66
4052 Basel, Switzerland

This is a reprint of articles from the Special Issue published online in the open access journal *Coatings* (ISSN 2079-6412) (available at: https://www.mdpi.com/journal/coatings/special_issues/Metal_Alloy).

For citation purposes, cite each article independently as indicated on the article page online and as indicated below:

LastName, A.A.; LastName, B.B.; LastName, C.C. Article Title. *Journal Name* **Year**, *Volume Number*, Page Range.

ISBN 978-3-0365-1010-1 (Hbk)
ISBN 978-3-0365-1011-8 (PDF)

Contents

About the Editors

Oriol Rius-Ayra (PhD) is a Postdoc researcher at the Universitat de Barcelona. His research is focused on the development of superhydrophobic surfaces on metals and alloys and studying their environmental applications, such as oil/water separation or the removal of microplastics.

Nuria Llorca-Isern is Professor at the Material Science and Engineering section of the Faculty of Chemistry at the Universitat de Barcelona. Her research is focused on surface modification technologies, powder metallurgy, and severe plastic deformation processes to achieve and enhance material properties to meet new material requirements from a sustainable perspective.

coatings

Editorial

Special Issue "Surface Modification of Metals and Alloys"

Oriol Rius-Ayra * and Nuria Llorca-Isern

CPCM, Department of Materials Science and Physical Chemistry, Faculty of Chemistry, Universitat de Barcelona, Martí i Franquès 1-11, 08028 Barcelona, Spain; nullorca@ub.edu
* Correspondence: oriolriusayra@ub.edu

Citation: Rius-Ayra, O.; Llorca-Isern, N. Special Issue "Surface Modification of Metals and Alloys". *Coatings* **2021**, *11*, 260. https://doi.org/10.3390/coatings11020260

Received: 17 February 2021
Accepted: 22 February 2021
Published: 23 February 2021

Recently surface modification has become necessary for the scientific community because of surface properties of new materials are usually inadequate in terms of wettability, adhesion, corrosion resistance or even drag reduction. In order to modify solid surfaces such as metals and alloys different treatments have been used to obtain a desired surface finish such as chemical vapor deposition, physical vapor deposition, chemical etching, electrodeposition or the application of non-equilibrium gaseous media, especially gaseous plasma. These treatments promote changes in roughness, hydrophobicity, biocompatibility or reactivity. Although such treatments have been studied extensively in past decades and actually commercialized, the exact mechanisms of interaction between reactive gaseous species and solid materials are still inadequately understood.

Moreover, for various reasons, it is difficult to find an alloy with a different surface behaviour from that of the bulk. A greater or more specific to extreme environments: resistance to corrosion and to wear, higher mechanical or fatigue resistance, hydrophobicity, oleophilicity, thermal (for low or high temperature exposure), magnetic, electrical or specific optic or light exposure behavior or to create biocompatibility or (bio)fouling or even their combined effect. In order to achieve and improve these properties in metals and alloys, we have to apply the strategy of surface modification based on a direct action on the metal or incorporating a coating that will provide these properties or functionalize its surface for complex requirements.

In general, the topics of interest range from newer approaches of conventional coatings technologies and thermomechanical processes, biocoatings, and surface modification for energy applications, catalysis, and nanomaterials to functionalization of metallic powder and additive manufactured metallic surfaces. All these improvements will be focused on developing successful engineering products and parts; some new strategies will also contribute to solving environmental issues.

This special issue provides recent trends in nanostructuring and functionalization of solid materials such as metals and alloys with the goal of improving their functional properties. Many different and innovative approaches that can be used to transform the metallic surface by means of physical, chemical, mechanical or biological characteristics providing different properties from the ones originally found on these surfaces allowing functionality for a given application and also for improving their properties.

Garfias Bulnes, Albaladejo Fuentes, Garcia Cano and Dosta [1] showed an analysis of the differences found in hard metal coatings produced by two high velocity thermal spray techniques, namely high velocity oxy-fuel (HVOF) and high velocity air-fuel (HVAF). Additionally, the effect of the metallic matrix and ceramic composition and the original carbide grain size on coating properties is compared to the most studied standard reference material sprayed by HVOF, WC-Co. For this evaluation, the physical properties of the coatings, including feedstock characteristics, porosity, thickness, roughness, hardness, and phase composition were investigated. Several characterization methods were used for this purpose: optical microscopy (OM), scanning electronic microscopy (SEM), Energy-dispersive X-ray spectroscopy (EDS), and X-ray Diffraction (XRD), among others. The final performance (abrasive wear and corrosion resistance) shown by the coatings obtained by these two methodologies was also analyzed. Thus, the abrasive wear resistance was analyzed by the rubber-wheel test, while the corrosion resistance was characterized with electrochemical methods. The characterization results obtained clearly showed that the coatings exhibit different microstructures according to feedstock powder characteristics (carbide grain size and/or composition) and the thermal spray process used for its deposition. Thus, the incorporation of WB to the cermet composition led to a high hardness coating, and the complementary hardness and toughness of the WC-Co coatings justify its better abrasion resistance. The presence of Ni on the metal matrix increases the free corrosion potential of the coating to nobler region.

Sendino, Gardon, Lartategui, Martínez and Lamikiz [2] showed the manufacture of multiple parts in the Laser Powder Bed Fusion (L-PBF) process. The main advantage is that the entire working volume of the machine is used and a greater number of parts is obtained, thus reducing inert gas volume, raw powder consumption, and manufacturing time. However, one of the main disadvantages of this method is the possible differences in quality and surface finish of the different parts manufactured on the same platform depending on their orientation and location, even if they are manufactured with the same process parameters and raw powder material. Throughout this study, these surface quality differences were studied, focusing on the variation of the surface roughness with the angle of incidence of the laser with respect to the platform. First, a characterization test was carried out to understand the behavior of the laser in the different areas of the platform. Then, the surface roughness, microstructure, and minimum thickness of vertical walls were analyzed in the different areas of the platform. These results were related to the angle of incidence of the laser. As it was observed, the laser is completely perpendicular only in the center of the platform, whilst at the border of the platform, due to the incidence angle; it melts an elliptical area, which affects the roughness and thickness of the manufactured part. The roughness increases from values of Sa = 5.489 μm in the central part of the platform to 27.473 μm at the outer borders while the thickness of the manufactured thin walls increases around 40 μm.

Pezzato, Settimi, Cerchier, Gennari, Dabalà and Brunelli [3] reported a Plasma Electrolytic Oxidation (PEO) as a surface treatment and produced a thick oxide films on the surface of metals. In their work, PEO coatings were obtained on zinc-aluminized (ZA) carbon steel using a solution containing sodium silicate and potassium hydroxide as electrolyte, and working with high current densities and short treatment times in Direct Current (DC) mode. The thickness of the coating, as well as the surface morphology, were strongly influenced by the process parameters, with different dissolution grades of the ZA layer depending on the current density and treatment time. A compromise between thickness and porosity of the coating was found with low current density/long treatment time or high current density/short treatment time. The PEO layer was mainly composed of aluminum oxides and silicon compounds. The corrosion resistance increased remarkably in the samples with the PEO coating. These PEO coated samples are suitable for sealing treatments that further increase their corrosion properties. They will be also an ideal substrate for commercial painting, assuring improved mechanical adhesion and protection even in the presence of damages

Li, Cui, Zhu, Narasimalu and Dong [4] presented a fluoropolyurethane-encapsulated process that was designed to rapidly fabricate low-flow resistance surfaces on the zinc substrate. For the further enhancement of the drag-reduction effect, Cu^{2+}-assisted chemical etching was introduced during the fabrication process. The resulting surface morphology, wettability and flow-resistance properties in a microchannel were also studied. Zinc substrate with a micro-nanoscale roughness obtained by Cu^{2+}-assisted nitric acid etching showed superhydrophilic characteristics. However, after the etched zinc substrate was encapsulated with fluoropolyurethane, the superhydrophobic wettability can be obtained with a contact angle of $154.8^\circ \pm 2.5^\circ$ and a rolling angle of less than 10°. As this newly fabricated surface was placed into a non-standard design microchannel, it was found that with the increase of Reynolds number, the drag-reduction rate of the superhydrophobic surface remained almost unchanged at 4.0% compared with the original zinc substrate. Furthermore, the prepared superhydrophobic surfaces exhibited outstanding reliability in most liquids.

Rius-Ayra, Fiestas-Paradela and LLorca-Isern [5] reported a straightforward method for water-harvesting based on modifications of the surface wettability by using the electrodeposition process of magnesium chloride, lauric acid revealing a superhydrophobic surface (155°). Morphological characterization techniques showed the characteristic flower-like microstructurs combined with close packed nanoarrays that were vertically aligned in a non-linear cone morphology formed by dynamic templating of hydrogen bubbles. From a chemical point of view, magnesium laurate is responsible for the surface tension decrease. To determine the durability an abrasive paper test was carried out revealing high durability against this severe condition. The water-harvesting ability of the superhydrophobic surface was studied at 45° and 90° tilted samples. The capacity of the water to be efficiently harvested was found to be at 90° tilt under fog conditions. The use of green reactants associated with this hierarchical structure broadens a new scope for sustainable freshwater collection and it becomes an excellent example of a green solution.

References

1. Bulnes, A.G.; Fuentes, V.A.; Cano, I.G.; Dosta, S. Understanding the Influence of High Velocity Thermal Spray Techniques on the Properties of Different Anti-Wear WC-Based Coatings. *Coatings* **2020**, *10*, 1157. [CrossRef]
2. Sendino, S.; Gardon, M.; Lartategui, F.; Martinez, S.; Lamikiz, A. The Effect of the Laser Incidence Angle in the Surface of L-PBF Processed Parts. *Coatings* **2020**, *10*, 1024. [CrossRef]
3. Pezzato, L.; Settimi, A.G.; Cerchier, P.; Gennari, C.; Dabalà, M.; Brunelli, K. Microstructural and Corrosion Properties of PEO Coated Zinc-Aluminized (ZA) Steel. *Coatings* **2020**, *10*, 448. [CrossRef]
4. Li, Y.; Cui, Z.; Zhu, Q.; Narasimalu, S.; Dong, Z. Fabrication of Zinc Substrate Encapsulated by Fluoropolyurethane and Its Drag-Reduction Enhancement by Chemical Etching. *Coatings* **2020**, *10*, 377. [CrossRef]
5. Rius-Ayra, O.; Fiestas-Paradela, S.; Llorca-Isern, N. Non-Fluorinated, Sustainable, and Durable Superhydrophobic Microarrayed Surface for Water-Harvesting. *Coatings* **2020**, *10*, 314. [CrossRef]

coatings

MDPI

Article

Understanding the Influence of High Velocity Thermal Spray Techniques on the Properties of Different Anti-Wear WC-Based Coatings

Andrea Garfias Bulnes, Vicente Albaladejo Fuentes, Irene Garcia Cano * and Sergi Dosta

Thermal Spray Centre, CPT, Departament CMiQF, Universitat de Barcelona, 08028 Barcelona, Spain; agarfias.b@gmail.com (A.G.B.); valbaladejo@cptub.eu (V.A.F.); sdosta@cptub.eu (S.D.)

* Correspondence: igcano@cptub.eu

Received: 15 October 2020; Accepted: 24 November 2020; Published: 26 November 2020

Abstract: This work analyzes the differences found in hard metal coatings produced by two high velocity thermal spray techniques, namely high velocity oxy-fuel (HVOF) and high velocity air-fuel (HVAF). Additionally, the effect of the metallic matrix and ceramic composition and the original carbide grain size on coating properties is compared to the most studied standard reference material sprayed by HVOF, WC-Co. For this evaluation, the physical properties of the coatings, including feedstock characteristics, porosity, thickness, roughness, hardness, and phase composition were investigated. Several characterization methods were used for this purpose: optical microscopy (OM), scanning electronic microscopy (SEM), Energy-dispersive X-ray spectroscopy (EDS), and X-ray Diffraction (XRD), among others. The final performance (abrasive wear and corrosion resistance) shown by the coatings obtained by these two methodologies was also analyzed. Thus, the abrasive wear resistance was analyzed by the rubber-wheel test, while the corrosion resistance was characterized with electrochemical methods. The characterization results obtained clearly showed that the coatings exhibit different microstructures according to feedstock powder characteristics (carbide grain size and/or composition) and the thermal spray process used for its deposition. Thus, the incorporation of WB to the cermet composition led to a high hardness coating, and the complementary hardness and toughness of the WC-Co coatings justify its better abrasion resistance. The presence of Ni on the metal matrix increases the free corrosion potential of the coating to more noble region. However, the WC-Co coatings show a lower corrosion rate and hence a higher protective performance than the rest of the coatings.

Keywords: thermal spray; HVOF; HVAF; WC-based coatings; cermet materials; wear resistance

1. Introduction

In recent years, the technological demands of industry have exceeded the performance of traditional coatings. Therefore, new coatings that are capable of fulfilling these needs have attracted important attention.

One example of these materials are cemented carbides, which are commonly known as hard metals and represent a group of hard and resistant compounds. They are a mixture of metallic and ceramic particles that combine the hardness and resistance of carbides (WC, TiC, TaC) with the toughness and plasticity of a metallic binder (Co, Ni, Cr). The properties of the hard metal coatings depend mainly on its composition and on the physical characteristics of its raw materials, hence any variation in these two parameters represents a change in the final microstructure, making it possible to obtain final specific tailored properties for particular applications [1].

Specifically, tungsten carbides (WC) are commercially one of the most successful metallurgic products in industry because of their combination of physical, chemical, and mechanical properties, which make them useful for applications in cutting tools and machinery exposed to wear [2]. In this respect, different WC-based hard metals can be found on the market showing different properties according to the diverse metallic binder and metal/ceramic compositions, as mentioned above.

Among the different WC-based cermets, WC-Co (with cobalt percentage ranging from 5 wt.% to 25 wt.%) coatings remain the most studied materials for industrial applications. Because of the great cohesion between the tungsten carbide grains and the cobalt metallic binder, these are composite materials with a great hardness–toughness relation, which makes them one of the best wear resistant materials. Tungsten carbides with cobalt have excellent mechanical properties. However their poor corrosion resistance restricts the number of applications in which they may be used, especially in the chemical industry [3–5].

As an alternative, the cobalt metallic binder can be substituted by nickel and/or nickel alloys in order to enhance its anticorrosive and thermal crack resistance properties. Since cobalt is expensive and its accessibility is complicated, the use of nickel (WC-Ni-based) lowers the costs and facilitates the supplies [2]. With the partial or complete substitution of cobalt with nickel, a considerable improvement in resistance to corrosion while maintaining anti-wear properties is to be expected [2,6–8].

Another alternative for the improvement of the hard metal performance is the addition of tungsten boride to the matrix. The WB grains are hard ceramic particles that will likely improve the hardness of the coatings. It has been shown that the WB incorporation enhances the mechanical properties of the substrate, since it has a high thermal shock resistance, high conductivity, and high corrosion and wear resistance [9–11].

During the production processes of hard metals, the WC suffer decarburization and oxidation, reacting to form di-tungsten carbide (W_2C), metallic tungsten (W) and oxy carbides. These reactions are inconvenient for the enhancement of the mechanical properties of the coatings, thus the subproducts formation must be reduced by controlling the process conditions. It has been found that the addition of chromium (WC-Co–Cr) somehow modifies the decomposition of the WC grains, preventing decarburization, while improving the matrix binding [12–16].

The final coating properties depend not only on its composition, but also on its lamellar microstructure. The formation of this lamellar microstructure depends on three main factors: the feedstock characteristics (material, size and morphology), the process method and parameters, and the substrate [17].

The feedstock material characteristics depend on their manufacturing process. Other than chemical composition, important factors are the size distribution, morphology, porosity bulk density, and grain size of the particles. Understanding the characteristics of the powder is crucial for the interpretation of the properties of the coating. Denser feedstock particles lead to coatings with better mechanical properties [17].

On the other hand, one method to deposit these types of materials in the form of coatings is by thermal spraying (TS). In general terms, in TS, the material feedstock is heated up in the form of powder using a hot gas stream until the melting or semi-melting of the particles. Then, they are accelerated to high velocities towards a substrate, leading to the adhesion of the semi-molten particles to the surface. From the TS technology arise several techniques, driven by combustion or electric discharge. Flame thermal spray, high velocity oxygen/air fuel (HVOF/HVAF), and detonation spraying (D-Gun) are based on combustion, while wire arc and plasma are based on electrical energy [18].

Among all the TS processes, HVOF/HVAF (high velocity oxygen/air fuel) are the most extended for obtaining coatings made of cemented carbides [19]. In these two processes (HVOF and HVAF), a mixture of gases is introduced into a combustion chamber. Inside this chamber, a spark generates a combustion reaction that heats the flowing gases to high temperatures. As the gases exit the chamber, powder particles are injected either axially or radially into the hot gases stream. This new outlet stream (heated mixture of gases and powder particles) flows into a De Laval type nozzle with a convergent-divergent structure that drastically reduces the flow area, leading to a considerable increase in the mixture flow velocity. As the gas velocity is increased at the outlet of the nozzle, the powder particles are ejected at supersonic speeds until they collide with a substrate and produce a coating [20].

The general equipment arrangement of the HVOF and HVAF processes is the same, however some important differences, summarized in Table 1, are to be noted regarding the operation conditions. In the case of HVOF, the gas mixture that is introduced into the combustion chamber consists of a fuel species (either hydrogen, propylene, methane, propane or heptane), nitrogen (inert gas) and oxygen (oxidizing component). In contrast, in the HVAF process, the oxygen of the gas mixture is replaced by dry air. This difference in the gas mixture composition has an impact on the flame temperature during the spraying process. HVOF jets show higher temperatures, in the range of 2700 and 3100 °C depending on the stoichiometry of the mixture, whilst HVAF operates at lower temperatures between 1900 and 2000 °C due to the higher ratio of N_2 (inert component) in the gas flow.

Table 1. Operation conditions differences in HVOF and HVAF processes.

Variable	HVOF	HVAF
Gas mixture	Fuel, nitrogen and oxygen	Fuel, nitrogen and dry air
Flame Temperature	2700–3100 °C	1900–2000 °C
Pressure	<1 MPa	<2 MPa
Particle velocity	600–750 m/s	600–840 m/s
Oxidizing power	High	Low

The lower ratio of fuel/oxygen used in HVAF compared to HVOF, allows to introduce a higher total gas amount into the combustion chamber. Consequently, the gas flow inside the nozzle is choked, leading to an increase in pressure inside the equipment. When the pressured gas-particle mixture flows and expands through the convergent-divergent nozzle, this pressure decreases as the velocity increases continuously. Commonly, HVOF operates at pressures up to 1 MPa, that induce particle velocities from 600 to 750 m/s, whereas the HVAF spraying process can reach up to 2 MPa, leading to particle velocities in the range of 600–840 m/s [18].

An efficient way to control the flame temperature and the velocity of the particles is by decreasing the fuel/oxygen ratio of the gases, and by increasing the flow rate inside the chamber through the injection of inert components (such as N_2). The control of these two variables is important, because at higher temperatures and lower particle velocities (which is the case of the HVOF process), the powder particles are more susceptible to undergo decarburization, resulting in undesired phases such as W_2C and W [6]. In the case of HVAF, the powders do not suffer significant decarburization or oxidation, as they are subjected to lower temperatures and exit the equipment at higher velocities [21].

The final properties of the coatings not only depend on the TS process used, but on the feedstock material characteristics too. The chemical composition, particle size distribution, morphology, porosity bulk density, and grain size of the particles are important features that should be taken into account when spraying WC-based powders. Thus, understanding the characteristics of the powder is crucial for the interpretation of the properties of the coating. Denser feedstock particles lead to coatings with better mechanical properties [17].

Because of this dependence of the coatings properties on both the feedstock characteristics and thermal spray processes, comparative studies have gained a lot of interest over the last years. This work analyzes two different thermal spray methods for the production of hard metals, namely HVOF and HVAF, by comparing the most studied standard reference material, WC-Co, sprayed by HVOF, with (a) WC-cermet coatings with different metallic matrix/ceramic composition and (b) WC-cermet coatings with different carbide sizes. This comparison comprises a study of the physical properties that constitute the microstructure of the coatings (feedstock characteristics, porosity, thickness, roughness, hardness, and phase composition) and mechanical properties (wear and corrosion resistance).

2. Materials and Methods

2.1. Feedstock Powders

The sprayed coatings were made with six different commercial powders. A WC-Co powder (AMPERIT 518.074, Goslar, Germany) was selected as reference material for obtaining HVOF and HVAF coatings.

To reveal the effect of the metallic matrix and ceramic compositions, three other materials with micrometric WC-grain size were sprayed: WC-Co–Cr, WC-NiMoCrFeCo, and WC-WB-CoCr with the powders AMPERIT 556.074, AMPERIT-529.074, and AMPERIT-539.074, respectively.

The relevance of feedstock powder carbide grain size in the final coating performances was evaluated using two other coatings with nanometric WC-grain size were sprayed: WC-Co with the powder FUJIMI W555-20/5 (Kakamigahara, Japan) and WC-Co–Cr with FUJIMI W928-32/10.

Table 2 shows the commercial names of the powders along with their coating identification names for this work, in which the suffix -m is included in the label for powders produced by AMPERIT (micrometric WC grain size) or -n for powders produced by FUJIMI (nanometric WC grain size).

Table 2. Commercial Powder information and Coatings identification.

Coatings Identification	Commercial Name	Particle Size (μm)	Composition	TS Method
HVOF-Co-m *	AMPERIT 518	−45 + 15	WC-12Co	HVOF
HVAF-Co	FUJIMI W55520/5	−45 + 15	WC-12Co	HVAF
HVOF-Co-n *	FUJIMI W55520/5	−20 + 5	WC-12Co	HVOF
HVOF-Co–Cr-m *	AMPERIT 556	−45 + 15	WC-10Co-4Cr	HVOF
HVOF-Co–Cr-n *	FUJIMI W9283210	−32 + 10	WC-10Co-4Cr	HVOF
HVOF-Ni	AMPERIT 529	−45 + 15	WC-15NiMoCrFeCo	HVOF
HVAF-Ni	AMPERIT 529	−45 + 15	WC-15NiMoCrFeCo	HVAF
HVOF-WB	AMPERIT 539	−45 + 15	WC-30WB5Co5Cr	HVOF
HVAF-WB	AMPERIT 539	−45 + 15	WC-30WB5Co5Cr	HVAF

* In these powders, the acronym -m and -n indicates the presence of micrometric and nanometric grain size of carbides in the powders microstructures respectively.

2.2. Thermal Spraying Conditions

All powders were sprayed on flat low carbon steel substrates (100 mm × 20 mm × 5 mm) and cylindrical low carbon steel substrates (φ = 25.4 mm × h = 25.4 mm), previously grit blasted with alumina. All blasted substrates had a mean roughness (Ra) greater or equal than 4μm.

All the coatings prepared by HVOF and HVAF methods were deposited in 15 layers sprayed at a 500 mm/s rate and at a distance of 250 mm, except for powders WC-Ni and WC-WB in the HVOF process which were sprayed at a distance of 300 mm. For the HVOF procedure was used a DJH 2600 gun (found in CPT facilities, Sultzer Metco, Pfäffikon, Switzerland) with a gas flow of hydrogen, oxygen, and air of 635, 267, and 344 L/min, respectively.

2.3. Powder and Coating Characterization

The size of the particles of the powders was analyzed in duplicate with a laser diffraction particle sizing analyzer LS 13 320 Model Dry Powder System (Beckman Coulter Inc., Indianapolis, IN, USA).

The metallographic preparation of the coatings was made according to the standards ASTM E1920-03(2014) [22]. Cross-sectional surface characteristics and the thickness of the coatings were analyzed with the optical microscope Leica DMI500M (Wetzlar, Germany) according to the standard ASTM E3-11(2017) [23].

Roughness was analyzed with 10 measures per coating with the profilometer SJ-210 Mitutoyo (Aurora, IL, USA). The mean porosity value was obtained from 30 measures with the software ImageJ (Version: 1.50i) from images at 20× magnification taken with the optical microscope according to standard ASTM E2109-01(2014) [24].

A Phenom ProX Desktop SEM (Waltham, MA, USA) was used in order to compare the superficial characteristics of the transversal side of the coatings and the feedstock powders. Elemental microanalysis of the composition of the samples was carried out with the equipment Quantax EDS Bruker Nano (Billerica, MA, USA) integrated to a SEM JEOL5310 (Akishima, Japan).

In order to analyze the phases and crystal structure of the coatings, the XRD equipment PANalytical X'Pert PRO MPD (PANalytical, Almelo, The Netherlands)was used with a radiation of Cu Kα (λ = 1.5418 Å) from 5 until 100° 2θ with a 0.017° step, measuring 100 s per step.

Vickers hardness and toughness were assessed according to the standard ASTM E384-17 [25] and following the work described by Lima, et al. [26], respectively. For each coating, 10 indentations were made with a force of 0.3 and 1 kgF respectively, with Matzuzawa MTZ-alfa equipment.

The adhesion of the coating to the steel substrate was measured with equipment SERVOSIS MCH-102ME (Madrid, Spain) according to the standard ASTM C633-13 [27]. Three samples of each coating were adhered to a non-sprayed counterpart with the adhesive HTK Ultrabond 100 (Hamburg, German). A tensile test was performed at a rate of 0.01 mm/s until cohesive, adhesive, or glue failure.

For the wear resistance to abrasion analysis, the Rubber Wheel method was tested with the tailored equipment CM4 OL-2000 (CM4, Cervelló, Barcelona, Spain) according to the standard ASTM G65-16e1 [28]. The test was performed at a velocity of 139 rpm, with a force of 50 N, a 22.6 cm diameter wheel, and Ottawa silica sand as the abrasive agent.

The corrosion resistance of the coatings was characterized with electrochemical experiments in 80 mL solution of NaCl 3.5%. The polarization curve was tested using the equipment VSP (Biologic Science Instruments Seyssinet-Pariset, Auvergne-Rhone-Alpes, France) at normal conditions in a potential range of 100 to +350 mV with respect to open circuit potential (E_{opc}) over 1 cm^2 sample surface. The corrosion potential (E_{corr}) and corrosion current (I_{corr}) were calculated with the software EC-Lab (V10.44). E_{corr} was obtained from a Tafel Fit extrapolation, while I_{corr} was calculated according to the Stern–Geary Equation (1).

$$I_{corr} = \frac{\beta_a \cdot \beta_c}{(2.3 \cdot (\beta_a + \beta_c))R_p} \quad (1)$$

where β_a and β_c are the anodic and cathodic currents, respectively, and R_p is the polarization resistance.

3. Results and Discussion

3.1. Powder Characterization

In thermal spraying technology, it is accepted that the microstructure of the coatings depends considerably on the characteristics of the powder particles at the point of collision with the substrate. In the HVOF/HVAF processes, the residence time of the particles inside the equipment, although minimal due to the high velocities (which depend on the gases flow rate) of the outlet flow of the gases, controls the heat transfer between the hot gases and the feedstock material and consequently determines its temperature at the point of impact with the substrate. At this point, the microstructure of the coating is influenced by the processes of deformation and solidification, caused by the high velocities and temperatures of the particles. These two factors are highly dependent on the size distribution of the powder particles, which is typically polydisperse. As a result, the particles of different sizes experience different momentum and thermal inertias, affecting the physical states of the particles during flight time and impact with the substrate [29,30]. Therefore, it is important to study the size distribution of the feedstock and adjust the process parameters accordingly. Hence, the characteristics and size distribution of the powders played an important role in the coatings microstructure, especially in terms of porosity and roughness.

The statistical size distribution of the particles and the size distribution curves for all the powders under study in this work are reported in Table 3 and Figure 1, respectively. This comparison shows that the powders mainly resulted to have a homogeneous distribution, as expected for commercial feedstock materials. Nonetheless, some differences can be found regarding powder properties depending on the feedstock material composition. Thus, WC-Co (micro and nanometric) revealed a relevant difference in their mean particle size of these powders, which shows as the particles of the WC-Co-m powder are larger than that of the WC-Co-n material. In addition, the nanometric powder showed a narrower size distribution than the micrometric one. No relevant differences were observed between powders WC-Co–Cr-m and WC-Co–Cr-n, which resulted in almost equivalent distributions with similar mean particle size values, as it can be seen in Figure 1b. On the contrary, the powders WC-WB and WC-Ni presented heterogeneous distributions, where the former one showed a distorted distribution with a bump, and the last one a distribution with two groups of particles with different sizes, resulting in the highest mean size distribution of all powders, with 80% of the particles with a diameter of between 20 and 61 µm.

Table 3. Particles size distribution.

Powder	d10 (µm) *	d90 (µm) *	Mean Particle Size (µm)
WC-Co-m	25	46	34
WC-Co-n	9	24	16
WC-Co–Cr-m	12	27	19
WC-Co–Cr-n	12	33	21
WC-Ni	20	61	35
WC-WB	14	48	27

* The particle size value in a dXX column indicates that the portion in percentage of particles with diameters smaller than that is XX%.

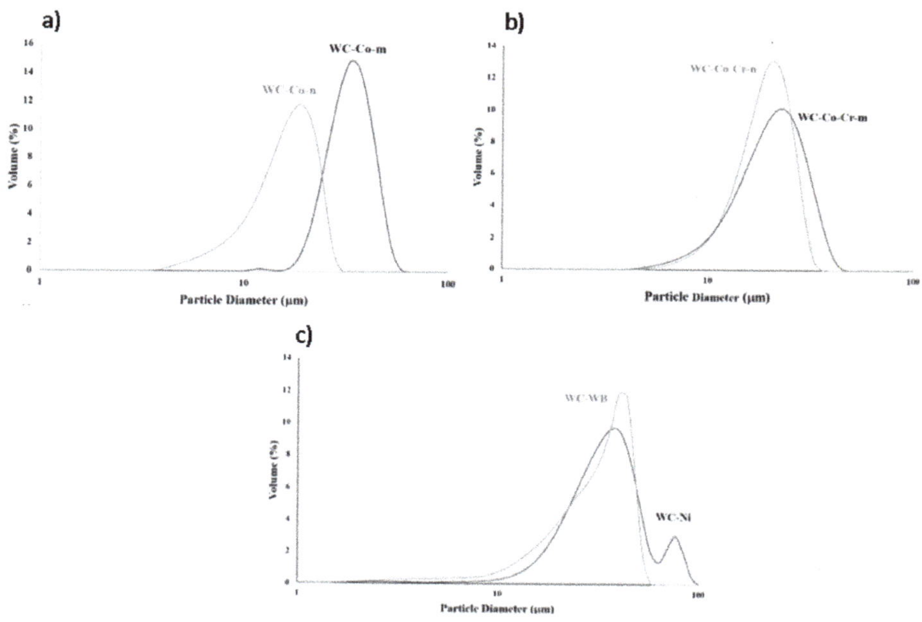

Figure 1. Distribution size of the powders WC-Co-n and WC-Co-m (**a**), WC-Co–Cr-n and WC-Co–Cr-m and (**b**) WC-WB and WC-Ni (**c**).

The spherical or irregular morphology of the powders and their microstructure was inspected by SEM. Figure 2 shows only the images recorded for the most representative powders (WC-Co-m, WC-Co-n, and WC-Co–Cr-m) for the sake of brevity. SEM images confirmed that all powders were composed of agglomerated particles and had a spherical morphology and were composed of WC grains embedded in a metallic binder: Co, Co–Cr (for WC-Co–Cr and WC-WB-Co–Cr powders), or NiMoCrFeCo, depending on the composition of the powder. Consistent with the particle size distribution results (Table 3), the free surface images showed the different sized particles of the powders. For instance, it is particularly interesting to observed the size diversity in the WC-Co-n powder (Figure 2c).

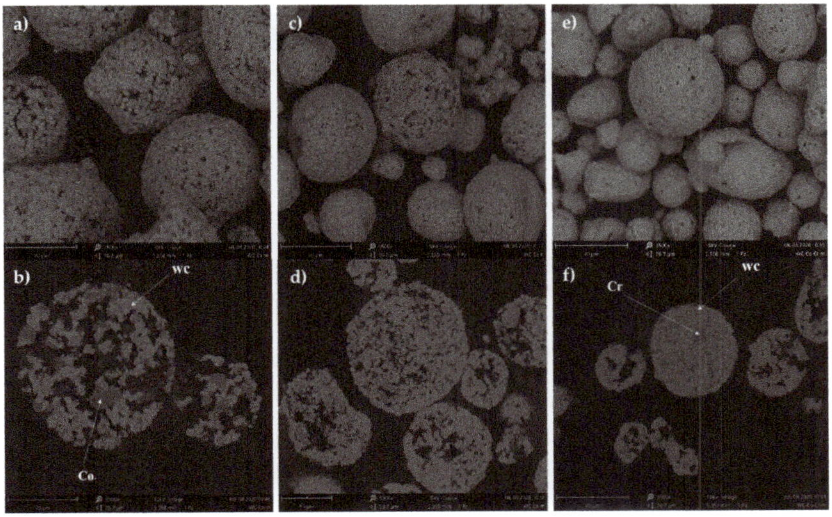

Figure 2. SEM images of the (**a**,**b**) WC-Co-m, (**c**,**d**) WC-Co-n and (**e**,**f**) WC-Co–Cr-m powders.

In contrast to the free surface images, that show the size difference of the agglomerated particles between all powders, in the cross-sectional images, the microstructure of the WC-based particles can be observed and analyzed [31]. Thus, the evaluation of the surface and core porosity level of the powders revealed that the WC-Co-m were slightly more porous than the WC-Co-n and WC-Co–Cr-n feedstock materials. Even though they are not shown, WC-Co–Cr-m powder presented similar porosity as WC-Co-m whilst powders WC-WB and WC-Ni were highly porous.

In order to identify the elements present in the powder composition, an EDS mapping analysis of the particles was performed and the images of this analysis are shown in Supplementary Figures S1–S3 of the supplementary information file of this manuscript. The results allow conclude that the dark regions of the particles microstructure (shown in Figure 2b,f) corresponded to the metallic binder of each powder, while the bright regions were ascribed to WC and/or WB grains. In this respect, in the powders with composition WC-Co, it was clearly identified that the metal matrix contained pure Co. Additionally, comparing their microstructure, in Figure 2b,d is confirmed that the powders labelled as micrometric showed higher size of the WC grains compared to the ones labelled as nanometric. Finally, in the case of the powders WC-Co–Cr and WC-WB, Cr predominated in the composition of the dark regions. In the same way, powder WC-Ni was mainly composed of Cr, Ni, and Mo.

For corroborating the presence of these phases, the XRD diffractograms of the powders are shown in Figure 3 According to the patterns, all the powders were found to contain only WC as a ceramic component and pure phases (Co, Cr, Ni, Mo) as the metallic binder.

In summary, the comparative analysis of the powders characteristics allowed concluding that all the powders are composed of WC grains embedded in a metallic matrix formed by metallic pure phases. WC-Co-m and WC-Co–Cr-m are composed of larger particles showing higher porosity than their respective nanometric powders. In addition, WC-Ni and WC-WB powders showed a broad size distribution, with two peaks in the WC-Ni case. These powders showed a highly porous microstructure.

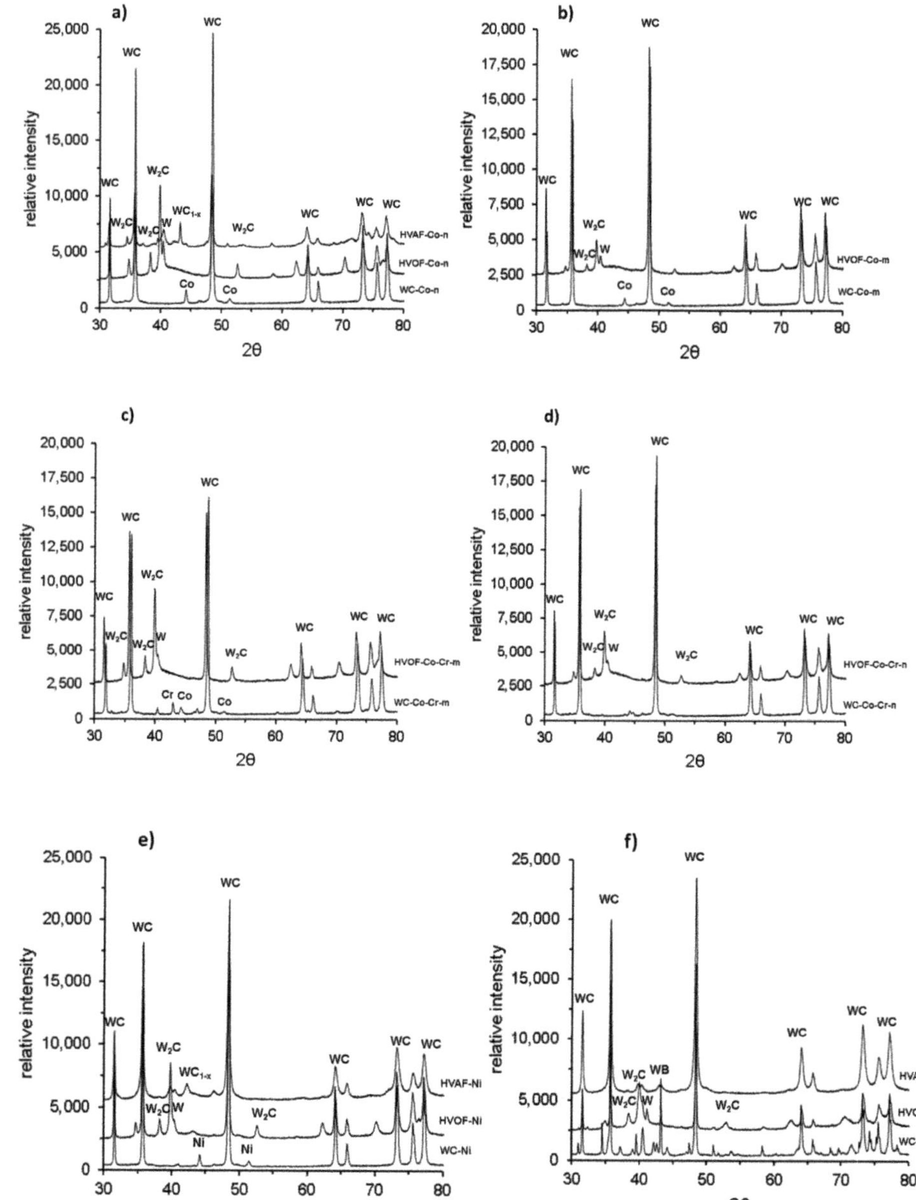

Figure 3. XRD of the powders and HVOF/HVAF coatings obtained using: (**a**) WC-Co-n, (**b**) WC-Co-m, (**c**) WC-Co–Cr-m, (**d**) WC-Co–Cr-n, (**e**) WC-Ni and (**f**) WC-WB feedstock materials.

3.2. Microstructure of the TS Coatings

It is worth remembering that the objective of this work was to compare the final coatings properties regarding the thermal spray process and composition and particle characteristics of feedstock materials. For this reason, the same spraying parameters were used in HVOF and HVAF processes respectively, for all the powders except for powders WC-Ni and WC-WB, which were sprayed at a higher distance in order to minimize high porosity values.

As expected, the coatings exhibited different microstructures according to their grain size, composition and thermal spray process used. Figure 4 shows representative polished cross-sectional recorded by optical microscopy (OM) (left) and SEM (right) images of the coatings, while Table 4 shows their thickness, porosity, roughness, and hardness values.

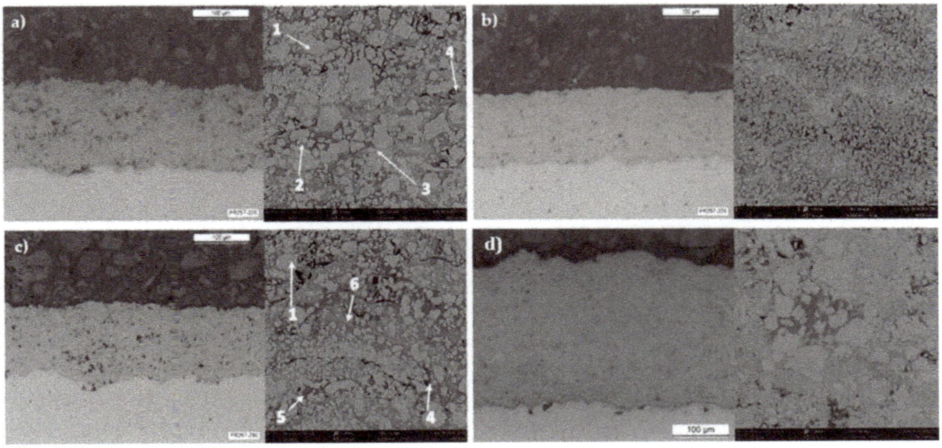

Figure 4. Cross-sectional OM (left at 20× magnification) and SEM (right) images of the coatings (**a**) HVOF-Co-m (9100×), (**b**) HVOF-Co-n (16,000×), (**c**) HVOF-Ni (12,000×), (**d**) HVAF-Ni (12,000×). Labels: 1 = WC, 2 = Co, 3 = Co + W_2C, 4 = pores, 5 = Ni + Cr + Mo and 6 = Ni + Cr + Mo + W_2C.

Table 4. Results of coating microstructure and physical properties.

Coating	Thickness (μm)	Porosity (%)	Roughness		Hardness (HV_{300})	Tensile Strength (MPa)
			Ra (μm)	Rz (μm)		
HVOF-Co-m	133.9 ± 9.2	2 ± 0.7	4.0 ± 0.4	21.5 ± 1.9	1580 ± 183	>65
HVOF-Co-n	120.7 ± 10.6	<1	2.3 ± 0.3	12.3 ± 1.0	1605 ± 157	>65
HVAF-Co-n	230.8 ± 7.7	<1	2.6 ± 0.2	14.6 ± 1.1	1669 ± 123	50.3 ± 6.0
HVOF-Co-Cr-m	118.0 ± 7.1	1.3 ± 0.6	2.4 ± 0.1	13.2 ± 0.9	1662 ± 141	>65
HVOF-Co-Cr-n	113.3 ± 7.2	<1	2.6 ± 0.3	14.4 ± 1.3	1630 ± 104	>65
HVOF-Ni	118.9 ± 11.0	1.8 ± 0.7	4.1 ± 0.3	23.1 ± 1.3	1508 ± 103	>65
HVAF-Ni	211.2 ± 5.3	<1	4.5 ± 0.3	24.8 ± 1.9	1474 ± 173	60.5 ± 8.0
HVOF-WB	117.1 ± 8.5	1.8 ± 0.6	5.3 ± 0.4	30 ± 3.1	1709 ± 142	>65
HVAF-WB	153.8 ± 9.0	9.7 ± 2.5	8.2 ± 0.6	43.6 ± 3.6	950 ± 104	48.2 ± 7.0

In general, from the results included in Table 4, it can be concluded that the coatings sprayed by HVAF were thicker, less porous and more homogeneous than the HVOF ones. Thus, the coatings sprayed by HVAF showed thickness values between 230 and 150 μm, against values between 133 and 113 μm. With the exception of the coating HVAF-WB, with a porosity value higher than 9%, all coatings had low porosity with values between <1% and 2%. In the case of the roughness of the coatings, it may be suggested that this is influenced by the cermet composition. In this respect, WC-Co coatings showed a smother surface than WC-Ni and WC-WB coatings. No clear differences are observed in the roughness values measured for the same coating compositions but obtained by HVOF or HVAF, with the exception of the coating HVAF-WB which presented a mean roughness Ra of 8.2 ± 0.6 μm, while the other HVOF coating presented Ra value of 5.3 ± 0.4 μm.

These data were found to be consistent with the porosity observed in the free surface and cross-sectional SEM images of the powders, as the powder used for preparing this coating also showed a high porosity. However, the coating porosity is not only dependent on the powders characteristics, but also on the final particle temperature during spraying process. When particles are heated up to higher temperatures, it is easier for them to melt on-flight before colliding with the substrate, resulting in denser coatings. Thus, despite the fact that WC-Co-m, WC-Co–Cr-m, and WC-Ni powders were very porous materials, the high density shown by their respective coatings allow for the conclusion that the spraying parameters used were aggressive enough for melting the particles.

Cross section images show that the particles that build up the coatings are strongly bonded to each other, since no cracks nor delamination areas were observed. Furthermore, the SEM images at high magnifications made it possible to observe the different phases in the microstructure of the coatings. All of them showed a homogenous distribution of the WC grains embedded into a darker metallic matrix along with some pores that might have been formed between particle splats. In Figure 4a,b the differences in the microstructure are clearly shown, especially regarding the WC grain size between micrometric and nanometric coatings. The WC grain size proved to be in the range of 1–4 μm for micrometric powders WC-Co-m and WC-Co–Cr-m, and 300–500 nm for nanometric powders WC-Co-n and WC-Co–Cr-n. Likewise powders WC-Ni and WC-WB had WC grains with sizes between <1–3 μm and <1–5 μm, respectively.

Looking at the coatings microstructure, it can be observed that those sprayed by HVOF showed some distinguished regions (darker than the WC grains), which might be attributed either to the metallic matrix or to the presence of undesired phases, difficultly distinguished in HVAF coatings. It is widely accepted that, in the thermal spraying methodology, the phase composition of a coating (WC-grains, metallic matrix and undesired phases) strongly depends on the flame temperature achieved. In this respect, the HVOF process leads to hotter and more oxidizing flame conditions than HVAF technique, which may explain these differences. In order to better identify these regions, the powders and coatings were further analyzed by EDS (Supplementary Figures S4–S6). This analysis confirmed the regions ascribed to the WC grains, W_2C, and metallic matrix, as indicated in the labels of Figure 4a,c. The identification of W_2C (lighter continuous regions surrounding WC grains with rounded shaped) allows concluding that certain processes of dissolution and decarburization of the WC grains occurred during spraying procedure. At high HVOF flame temperatures, the mechanism of decarburization starts with the melting of the metallic binder, since it has a lower melting point than the WC carbides. Afterwards, the WC grains begin to dissolve in the molten metal and, at the gas-liquid interface, carbon is released by its reaction with the oxygen of the gases. At the point of impact with the substrate, the particles are rapidly quenched and the dissolved WC grains become supersaturated, resulting in the formation of W_2C [32]. In contrast, the W_2C phase was not clearly observed in the microstructure of the HVAF coatings.

A microstructure comparison between HVOF and HVAF coatings can be observed in Figure 4c,d. In these images, it is shown that in the microstructure of the coating HVOF-Ni has a high presence of these lighter regions between different sized rounded grains, whereas coating HVAF-Ni only showed irregular grains embedded in the metallic matrix.

The XRD patterns of all the coatings were recorded for further support of the conclusion drawn from the visual inspection (Figure 3). As expected, all the coatings were composed mainly of the phases also identified in their respective powders, that is, WC as ceramic phase and their correspondent metallic pure phases as matrix. In agreement with previous results, HVAF coatings showed little or none undesired phases, while some HVOF coatings suffered relevant decarburization. Thus, in the case of the HVOF coatings patterns, the presence of W_2C and W phases indicated that decarburization occurred. As above mentioned, since the HVOF process operates at higher temperatures, the coatings sprayed by this method were more likely to form undesired phases. This is clearly observed in coatings HVOF-Co-n and HVOF-Co–Cr-n, as they were the ones that showed more content of W_2C and W. Among all the HVOF coatings, the HVOF-Co-m was the one that underwent less decarburization, since the sprayed conditions (used for all coatings) were the optimal ones for the size distribution of this conventional powder. For this reason, the nanostructured coating HVOF-Co-n showed more W_2C phases in comparison to the HVOF-Co-m one, which can be understood regarding the smaller WC grain size for the nanometric powder that facilitates the dissolution of WC original grains and their final decarburization during fast solidification. Another factor that affected the presence of the undesired phases is the binder composition. As mentioned before, the addition of Cr to the matrix is known to prevent the WC grains decarburization. This explains that the XRD patterns of both coatings HVOF-Co–Cr-n and HVOF-Co–Cr-m feature few differences between them in terms of W_2C presence and, for hence, it points out that less decarburization occurs than for the coating HVOF-Co-n. Coatings HVOF-Ni and HVOF-WB showed little decarburization, attributed to their composition and elevated mean particle size, as these bigger particles lightly experienced the elevated temperatures of the process.

Compared to HVOF methodology, during HVAF processes the powders are subjected to lower temperatures, therefore, except for the coating HVAF-Co-n which presented some W_2C, WC_{1-x} and W phases, most HVAF sprayed coatings showed practically no decarburization, meaning that the peaks corresponded in great part only to WC.

Finally, it is worth noting that the coatings HVOF-Co-n, HVOF-Co–Cr-m and HVOF-Co–Cr-n revealed one bump between 40°–45° indicating the presence of amorphous/nanochrystalline phases that were formed during the dissolution of the WC grains [15,33].

Thus, comparing the general microstructure of the coatings, it can be concluded that the ones sprayed by HVAF were thicker, less porous, and more homogeneous than the HVOF ones as they presented less roughness and practically no undesired phases due to the conservation of the WC grains during the TS process.

3.3. Mechanical and Anti-Wear Properties of the Coatings

The mechanical properties of the coatings are defined by its microstructure and composition. For instance, hardness values depend on several factors based on the microstructure of the coatings: (i) it increases as a function of cohesion between particles and coating porosity [34,35]; (ii) decarburization degree has an important impact on hardness since the higher amount of W_2C, the higher the coating hardness [36]; (iii) it depends on the WC grain size, as a small carbide size leads to a high hardness [37]; and (iv) the presence of an amorphous/nanocrystalline phase results in an increased of cohesion between the carbide grain and the metallic matrix, that leads to an improvement of the hardness of the coating [9].

Taking into account the previous discussion about the coatings microstructure, it would be expected that HVOF coatings showed higher hardness than the HVAF coatings because of the presence of W_2C in their composition. However, it becomes difficult to draw this conclusion looking at the Vickers hardness values (0.3 kgF) shown in Table 4. In general, these data revealed that all the coatings featured very high hardness, above 1500 $HV_{0.3}$, independently of the TS process used for their preparation. This result suggests that the higher degree of amorphous/nanocrystalline phases identified in the HVAF coatings microstructure plays a key role and enhanced the cohesion between the carbide grains and the matrix, and as a result, increased the hardness of the coating.

Making a comparison based only on composition, both HVAF and HVOF coatings presented roughly the same trend in terms of hardness values. The coating with the WB in the ceramic phase resulted to have the greater hardness (this is not applicable for coating HVAF-WB since it showed very high porosity), which is followed, in order, by coatings with the binders Co–Cr, Co, and Ni-alloy In the case of the micrometric and nanometric based coatings, the coating HVOF-Co-n resulted slightly harder than the conventional coating HVOF-Co-m, as the first one showed to have more W_2C and phases than the former one. For this same reason, Co–Cr based coatings presented an inverse trend: the coating HVOF-Co–Cr-m resulted slighlty harder than the HVOF-Co–Cr-n.

Usually, high coating hardness is related to a decrease in its toughness, nonetheless hard metals have shown to maintain high toughness when having high hardness values [38]. This unique combination of properties makes them excellent materials for applications that require high resistance to wear. Toughness fracture resistance mainly depends on two factors in this kind of coating, namely the amount and characteristics of the binder phase present and the grain size of the carbide phase [39]. It is worth indicating that an attempt was made to measure the toughness of these coatings following the calculation method presented in the work of Lima et al. [22]. Nevertheless, it was not possible to obtain the toughness values since the maximum indentation force of the equipment was not strong enough to produce the adequate length of the cracks needed in order to fulfill one of the restrictions of the method. Despite the fact that the toughness was not analytically calculated, it may be suggested that all coatings presented very high fracture toughness due to the heavy load necessary for crack formation and propagation. In addition, this observation would point out that, in these HVOF and HVAF coatings, there exists a high cohesion strength among the splats forming the coating microstructure.

Another important property of hard metals is their adhesion with the substrate surface, which determines the actual utility of the coating. Nowadays, it is considered that coatings prepared by conventional TS techniques are mechanically anchored to the substrate. Thus, a perfectly controlled cleaning and grit blasting process of the substrate has a strong influence on the adhesion strength at the interface. However, the degree of fusion of the particles but also the interfacial bonding within the lamellar microstructure, which integrates the pores size and distribution [17] will be also crucial for a proper coating performance in terms of adhesion.

The adhesion strength of the coatings was measured by means of tensile tests and the results obtained are included in Table 4. In the case of the HVAF coatings, HVAF-Ni, HVAF-Co, and HVAF-WB failed by adhesion of the coating–substrate interface at 60.5 ± 8.0 MPa, 50.3 ± 6.0 MPa, and 48.2 ± 7.0 MPa values, respectively. This kind of failure indicated that, at these tensile strength values, the coating was not strong enough to maintain its bond with the substrate. Moreover, from the analysis of these values, it can be concluded that the metallic binder has a relevant role in the adhesion resistance of the coatings, since the HVAF coating showing the highest adhesion strength is that with the highest metal/ceramic ratio and also with the lowest melting point, HVAF-Ni.

On the contrary, all the HVOF coating were able to withstand a higher tensile force than that withstand by the adhesive, exceeding the 65 MPa. In such a case, this is consistent with the difference in flame temperature of both processes, and the high flame temperature reached in the HVOF process allows the particles to fuse together and leads to stronger bonds within the microstructure.

It is important to highlight that the high adhesion strength values reported for all the coatings is in agreement with previous analysis, and supports the suggestion that there is high cohesion between the particles in coating.

Rubber wheel experiments were performed for the HVAF sprayed coatings (HVAF-Co, HVAF-Ni, HVAF-WB), their respective HVOF coatings (HVOF-Co-n, HVOF-Ni, HVOF-WB), and the conventional coating (HVOF-Co-m) in order to evaluate their wear resistance to abrasive conditions. Wear behavior of the coatings during the experiments are shown in Figure 5. In this plot, it is observed, within the first minutes of the test, that the wear rate significantly dropped due to the rapid decrease of the roughness of the coatings. The ones that presented higher roughness values (HVAF-Ni and HVAF-WB) showed a steeper drop in the wear rate curve in comparison to HVAF-Co and the steel substrate. In general, after the point where the surface evened out, the wear rate followed a flatter behavior for all the coatings, indicating the hardening of the coatings and the substrate.

Abrasive wear is the main mechanism of wear when the surface of two materials are facing each other with the presence of a third body in between. Abrasion then causes the removal of particles from the softest material. It has been shown that the wear resistance of the hard metals coatings is highly dependent on its composition, carbide grain size, porosity, toughness and hardness values [39]. The degree of abrasive wear is defined by the hardness difference between the abrasive particles and the counter material, hence higher hardness values are desired to enhance the wear resistance of the coatings and ensure the protection of the coated surface. In the case of the cermets, both hardness and cohesion between the carbides are key properties to improve the abrasive wear resistance. It is clear that the WC particles contribute to the overall hardness, while the metallic matrix defines the cohesion of the WC grains. Therefore, it is important to consider the size of the carbide grains, the binder composition and the thermal sprayed conditions when analyzing the wear resistance of the coatings [39].

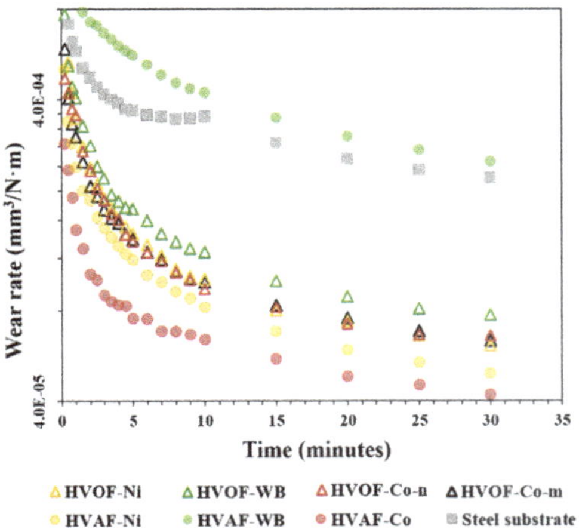

Figure 5. Wear rate results of the coatings HVAF-Co, HVAF-Ni, HVAF-WB, HVOF-Co-m, HVOF-Co-n, HVOF-Ni, HVOF-WB and the steel substrate.

In Table 5 are shown the volume loss (measured at the end of the experiment) and the wear rate (final value in Figure 5 for each coating) calculated for all the TS coatings. As can be observed from these data, the coatings HVAF-Co-n and HVAF-Ni resulted to have higher resistance to wear than the other TS coatings with a similar volume loss. This result can be understood if it is considered that these coatings showed low roughness, the absence of undesired phases, and one of the highest thickness and hardness values of the analyzed coatings. HVAF-WB coating experienced a high volume loss value of 37.0 mm^3, being even a greater value than the volume loss measured for the bare steel substrate, which can be explained because of its high porosity. For HVOF coatings, no clear differences in abrasive wear resistance were observed for the coatings HVOF-Co-m, HVOF-Co-n and HVOF-Ni as they showed practically identical wear rate values. In this case, HVOF-Co-n was expected to show the best performance (since nanostructured coatings are known to have enhanced mechanical properties), nevertheless this was not the case. As mentioned before, coating HVOF-Co-n suffered high decarburization, thus its abrasion resistance resulted degraded to slightly higher volume loss and wear rate values than coating HVOF-Co-m. In agreement with the results for HVAF technique, HVOF-WB coating also experienced a high volume loss, close to the value measure for the steel substrate (11.5 mm^3 and 11.7 mm^3, respectively). In this case, the HVOF-WB did not show a porosity that would explain the result, and for this reason, these volume loss and wear rates might be attributed to a poor cohesion between the WC/WB grains and the metallic matrix.

Table 5. Data collected from abrasive wear and corrosion resistance experiments.

Coating	Volume Loss (mm^3)	Wear Rate (mm^3/Nm)	E_{corr} (mV)	I_{corr} (μA)
HVAF-Co	6.2	4.21×10^{-5}	−447.7	0.8
HVAF-Ni	7.3	4.93×10^{-5}	−369.9	0.2
HVAF-WB	37.0	2.50×10^{-4}	−559.5	3.8
HVOF-Ni	9.0	6.09×10^{-5}	−480.9	1.4
HVOF-Co-m	9.4	6.35×10^{-5}	–	–
HVOF-Co-n	9.7	6.61×10^{-5}	−537.5	1.2
HVOF-WB	11.5	7.75×10^{-5}	−576.5	1.4
Steel substrate	11.7	2.20×10^{-4}	−726.5	1.4

The corrosion behavior of the HVAF and HVOF coatings was evaluated and the resulting polarization curves are shown in Figure 6a,b. The actual values of E_{corr} and I_{corr} inferred from the polarization curves and calculated with Equation (1), respectively, are shown in Table 5. According to this, a first differentiation can be made between coatings regarding the thermal spraying method used for its deposition, since, in general, the HVAF coatings show higher E_{corr} and lower I_{corr} than the HVOF layers. This result can be clearly related to the differences found in the microstructure of coatings. It was demonstrated that HVAF coatings show higher compactness than HVOF coatings, that is a lower porosity and the presence of voids between splats, which is widely accepted in literature as a responsible of their better corrosion behavior [40].

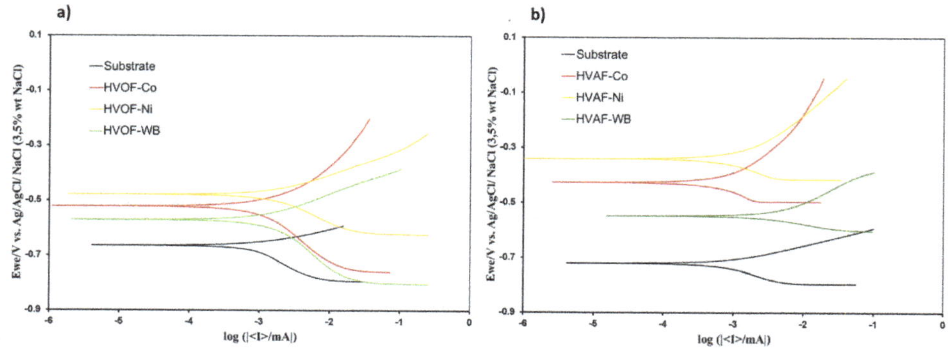

Figure 6. Polarization curves of the coatings by (**a**) HVOF and (**b**) HVAF and the steel substrate.

When subjecting a coated material to corrosive conditions, it is expected for the coating to prevent the effects of corrosion and to ensure the correct operation of the substrate. In hard metals the E_{corr} and I_{corr} values are intrinsic of the metallic matrix material. Taking into account the results shown in Figure 6, both HVOF and HVAF coatings can be classified as a function of their composition, from top to bottom, that is, from high to low E_{corr}, as follows: Ni-based (HVOF-Ni & HVAF-Ni), Co-based (HVOF-Co & HVAF-Co) and WB-based (HVOF-WB & HVAF-WB). These data can be justified since Ni is a more noble metal than Co or Co–Cr alloys.

On the other hand, the comparison of the I_{corr} values between HVOF-Co & HVOF-Ni and HVAF-Co and HVAF-Ni shows that these coatings present similar corrosion rates under open circuit potentials. Despite of this, when the potential is shifted to the anodic region, these coatings clearly show a different behavior since HVOF-Co and HVAF-Co coatings show a lower current intensity value than the Ni-based coatings over a potential region approximately of −340 and −185 mV, respectively. This data suggests that under polarization conditions, the Co-based coatings will corrode more slowly than the Ni-based coatings. Considering that this kind of coating is supposed to act as a protective barrier between the environment and the substrate, this allows for the conclusion that the HVOF-Co and HVAF-Co coatings will show higher corrosion protective performance than the rest of the coatings studied.

Finally, low coating porosity is mandatory for good corrosion protection. In the case of the HVAF-WB coating, which showed very high porosity, it still shows a certain protective ability since its E_{corr} is higher than that of the steel substrate. However, the high I_{corr} value measured for this material reveals that there is a severe rate of corrosion in this system which might be ascribed to steel corrosion, and hence it indicates that this is not a good protective coating.

4. Conclusions

The coatings exhibited different microstructures and mechanical properties according to their carbide size, composition, and the thermal spray process used. In general, all coatings showed a homogenous distribution of the WC grains embedded into a darker metallic matrix with strong cohesion between particles.

Different thermal spray methods, namely HVOF and HVAF, lead to coatings with different microstructures and mechanical properties. (a) Because of the higher flame temperature in HVOF, the coatings sprayed by this method were more likely to form undesired phases, resulting in the decarburization of the WC grains. (b) Due to the conservation of the WC grains during the TS process, HVAF coatings were thicker, less porous, and more homogeneous than the HVOF ones. (c) HVAF coatings were slightly more resistant to abrasive wear than HVOF coatings.

The metallic matrix, ceramic composition, and WC grain size were key factors in hardness and the mechanical properties of the coatings. (a) The coating with the WB matrix resulted to have the greater hardness, followed, in order, by coatings with the binders Co–Cr, Co, and Ni-alloy. (b) The WC-Co coating proved to be the one with higher resistance to abrasive wear and the one showing the higher corrosion protection to steel in NaCl media. (c) Coatings with a smaller grain lead to superior qualities of the coatings in terms of enhanced properties.

Supplementary Materials: The following are available online at http://www.mdpi.com/2079-6412/10/12/1157/s1, Figure S1: EDS mapping analysis of the WC-Co-m powder, (a) SEM image of the cross-section of the powder particle at 5000×, (b) Co-rich regions (blue) and (c) W-rich regions (red), Figure S2: EDS mapping analysis of the WC-Co-n powder, (a) SEM image of the cross-section of the powder particle at 5000×, (b) Co-rich regions (blue) and (c) W-rich regions (red), Figure S3: EDS mapping analysis of the WC-Co-Cr-n powder, (a) SEM image of the cross-section of the powder particle at 2000×, (b) Co-rich regions (blue), (c) W-rich regions (red) and (d) Cr-rich regions (yellow), Figure S4: EDS mapping analysis of the HVOF-Co-m coating, (a) SEM image of the cross-section of the coating at 5000×, (b) Co-rich regions (blue) and (c) W-rich regions (red), Figure S5: EDS mapping analysis of the HVOF-Co-n coating, (a) SEM image of the cross-section of the coating at 5000×, (b) Co-rich regions (blue) and (c) W-rich regions (red), Figure S6: EDS mapping analysis of the HVOF-Ni coating, (a) SEM image of the cross-section of the coating at 5000×, (b) Ni-rich regions (bright blue), (c) W-rich regions (red), (d) Cr-rich regions (yellow) and e) Mo-rich regions (pink).

Author Contributions: Conceptualization, A.G.B., V.A.F., I.G.C., and S.D.; methodology, A.G.B., V.A.F., and S.D.; software, A.G.B.; validation, A.G.B. and V.A.F.; formal analysis, A.G.B., V.A.F., I.G.C., and S.D.; investigation, A.G.B. and V.A.F.; writing—original draft preparation, A.G.B. and V.A.F.; writing—review and editing, I.G.C. and S.D.; supervision, I.G.C. and S.D.; project administration, V.A.F., I.G.C., and S.D.; funding acquisition, I.G.C. and S.D. All authors have read and agreed to the published version of the manuscript.

Funding: This research was funded by Programa Operatiu FEDER de Catalunya 2014–2020. Grant No. COMRDI16-1-0020.

Conflicts of Interest: The authors declare no conflict of interest.

References

1. Lassner, E.; Schubert, W.-D. *Tungsten*; Springer: Boston, MA, USA, 1999; ISBN 978-1-4613-7225-7.
2. Chang, S.H.; Chen, S.L. Characterization and properties of sintered WC-Co and WC-Ni-Fe hard metal alloys. *J. Alloys Compd.* **2014**, *585*, 407–413. [CrossRef]
3. Stewart, D.A.; Shipway, P.H.; McCartney, D.G. Abrasive wear behaviour of conventional and nanocomposite HVOF-sprayed WC-Co coatings. *Wear* **1999**, *225–229*, 789–798. [CrossRef]
4. Chivavibul, P.; Watanabe, M.; Kuroda, S.; Shinoda, K. Effects of carbide size and Co content on the microstructure and mechanical properties of HVOF-sprayed WC-Co coatings. *Surf. Coat. Technol.* **2007**, *202*, 509–521. [CrossRef]
5. He, J.; Schoenung, J.M. A review on nanostructured WC-Co coatings. *Surf. Coat. Technol.* **2002**, *157*, 72–79. [CrossRef]
6. Guilemany, J.M.; Sanchiz, I.; Mellor, B.G.; Llorca, N.; Miguel, J.R. Mechanical-property relationships of Co/WC and Co Ni Fe/WC hard metal alloys. *Int. J. Refract. Met. Hard Mater.* **1993**, *12*, 199–206. [CrossRef]
7. Guilemany, J.M.; Nutting, J.; Miguel, J.R.; Dong, Z. Microstructure characterization of WC-Ni coatings obtained by HVOF thermal spraying. *Scr. Met. Mater.* **1995**, *33*, 55–61. [CrossRef]
8. Testa, V.; Morelli, S.; Bolelli, G.; Benedetti, B.; Puddu, P.; Sassatelli, P.; Lusvarghi, L. Alternative metallic matrices for WC-based HVOF coatings. *Surf. Coat. Technol.* **2020**, *402*, 126308. [CrossRef]

9. Bouaricha, S.; Legoux, J.-G.; Marple, B.R. HVOF Coatings Properties of the Newly Thermal Spray Composition WC-WB-Co. In Proceedings of the International Thermal Spray Conference 2005 (ITSC 2005), Basel, Switzerland, 2–4 May 2005; pp. 981–986.
10. Brezinová, J.; Guzanová, A.; Draganovská, D.; Maruschak, P.O.; Landová, M. Study of selected properties of thermally sprayed coatings containing WC and WB hard particles. *Acta Mech. Autom.* **2016**, *10*, 296–299. [CrossRef]
11. Brezinová, J.; Landová, M.; Guzanová, A.; Dulebová, L.; Draganovská, D. Microstructure, wear behavior and corrosion resistance of WC-FeCrAl and WC-WB-Co coatings. *Metal* **2018**, *8*, 399. [CrossRef]
12. Murthy, J.K.N.; Rao, D.S.; Venkataraman, B. Effect of grinding on the erosion behaviour of a WC-Co-Cr coating deposited by HVOF and detonation gun spray processes. *Wear* **2001**, *249*, 592–600. [CrossRef]
13. Perry, J.M.; Neville, A.; Hodgkiess, T. A comparison of the corrosion behavior of WC-Co-Cr and WC-Co HVOF thermally sprayed coatings by in situ Atomic Force Microscopy (AFM). *J. Ther. Spray Technol.* **2002**, *11*, 536–541. [CrossRef]
14. Karimi, A.; Verdon, C.; Barbezat, G. Microstructure and hydroabrasive wear behaviour of high velocity oxy-fuel thermally sprayed WCCo(Cr) coatings. *Surf. Coat. Technol.* **1993**, *57*, 81–89. [CrossRef]
15. Pulsford, J.; Kamnis, S.; Murray, J.; Bai, M.; Hussain, T. Effect of particle and carbide grain sizes on a HVOAF WC-Co-Cr coating for the future application on internal surfaces: Microstructure and wear. *J. Ther. Spray Technol.* **2018**, *27*, 207–219. [CrossRef]
16. Wang, H.; Wang, X.; Song, X.; Liu, X.; Liu, X. Sliding wear behavior of nanostructured WC-Co-Cr coatings. *Appl. Surf. Sci.* **2015**, *355*, 453–460. [CrossRef]
17. Ang, A.S.M.; Berndt, C.C. A review of testing methods for thermal spray coatings. *Int. Mater. Rev.* **2014**, *59*, 179–223. [CrossRef]
18. Fauchais, P.L.; Heberlein, J.V.R.; Boulos, M.I. *Thermal Spray Fundamentals*; Springer: Boston, MA, USA, 2014; ISBN 978-0-387-28319-7.
19. Fauchais, P.; Vardelle, A. Thermal Sprayed Coatings Used Against Corrosion and Corrosive Wear. In *Advanced Plasma Spray Applications*; IntechOpen: London, UK, 2012; pp. 3–39. ISBN 978-953-51-0349-3.
20. Thorpe, M.L.; Richter, H.J. A pragmatic analysis and comparison of HVOF processes. *J. Ther. Spray Technol.* **1992**, *1*, 161–170. [CrossRef]
21. Hulka, I.; Şerban, V.A.; Niemi, K.; Vuoristo, P.; Wolf, J. Comparison of structure and wear properties of fine-structured WC-CoCr coatings deposited by HVOF and HVAF spraying processes. *Solid State Phenom.* **2012**, *188*, 422–427. [CrossRef]
22. ASTM E1920-03. *Standard Guide for Metallographic Preparation of Thermal Sprayed Coatings*; ASTM International: West Conshohocken, PA, USA, 2014.
23. ASTM E3-11. *Standard Guide for Preparation of Metallographic Specimens*; ASTM International: West Conshohocken, PA, USA, 2017.
24. ASTM E2109-01. *Standard Test Methods for Determining Area Percentage Porosity in Thermal Sprayed Coatings*; ASTM International: West Conshohocken, PA, USA, 2014.
25. ASTM E384-17. *Standard Test Method for Microindentation Hardness of Materials*; ASTM International: West Conshohocken, PA, USA, 2017.
26. Lima, M.M.; Godoy, C.; Avelar-Batista, J.C.; Modenesi, P.J. Toughness evaluation of HVOF WC-Co coatings using non-linear regression analysis. *Mater. Sci. Eng. A* **2003**, *357*, 337–345. [CrossRef]
27. ASTM C633-13. *Standard Test Method for Adhesion or Cohesion Strength of Thermal Spray Coatings*; ASTM International: West Conshohocken, PA, USA, 2017.
28. ASTM G65-16e1. *Standard Test Method for Measuring Abrasion Using the Dry Sand/Rubber Wheel Apparatus*; ASTM International: West Conshohocken, PA, USA, 2016.
29. Li, M.; Christofides, P.D. Modeling and control of high-velocity oxygen-fuel (HVOF) thermal spray: A tutorial review. *J. Ther. Spray Technol.* **2009**, *18*, 753–768. [CrossRef]
30. Li, M.; Christofides, P.D. Computational study of particle in-flight behavior in the HVOF thermal spray process. *Chem. Eng. Sci.* **2006**, *61*, 6540–6552. [CrossRef]
31. Guilemany, J.M.; Dosta, S.; Miguel, J.R. The enhancement of the properties of WC-Co HVOF coatings through the use of nanostructured and microstructured feedstock powders. *Surf. Coat. Technol.* **2006**, *201*, 1180–1190. [CrossRef]

32. Kear, B.; Skandan, G.; Sadangi, R. Factors controlling decarburization in HVOF sprayed nano-WC/Co hardcoatings. *Scr. Mater.* **2001**, *44*, 1703–1707. [CrossRef]
33. Magnani, M.; Suegama, P.H.; Espallargas, N.; Dosta, S.; Fugivara, C.S.; Guilemany, J.M.; Benedetti, A.V. Influence of HVOF parameters on the corrosion and wear resistance of WC-Co coatings sprayed on AA7050 T7. *Surf. Coat. Technol.* **2008**, *202*, 4746–4757. [CrossRef]
34. Celik, E.; Culha, O.; Uyulgan, B.; Ak Azem, N.F.; Ozdemir, I.; Turk, A. Assessment of microstructural and mechanical properties of HVOF sprayed WC-based cermet coatings for a roller cylinder. *Surf. Coat. Technol.* **2006**, *200*, 4320–4328. [CrossRef]
35. Yang, Q.; Senda, T.; Ohmori, A. Effect of carbide grain size on microstructure and sliding wear behavior of HVOF-sprayed WC-12% Co coatings. *Wear* **2003**, *254*, 23–34. [CrossRef]
36. Usmani, S.; Sampath, S.; Houck, D.L.; Lee, D. Effect of carbide grain size on the sliding and abrasive wear behavior of thermally sprayed WC-Co coatings. *Tribol. Trans.* **1997**, *40*, 470–478. [CrossRef]
37. Ettmayer, P. Hardmetals and cermets. *Annu. Rev. Mater. Sci.* **1989**, *19*, 145–164. [CrossRef]
38. Gahr, K.-H.Z. (Ed.) Chapter 2 Microstructure and Mechanical Properties of Materials. In *Microstructure and Wear of Materials*; Elsevier: Amsterdam, The Netherlands, 1987; Volume 10, pp. 8–47, ISBN 0167-8922.
39. Liao, H.; Normand, B.; Coddet, C. Influence of coating microstructure on the abrasive wear resistance of WC/Co cermet coatings. *Surf. Coat. Technol.* **2000**, *124*, 235–242. [CrossRef]
40. Liu, Y.; Liu, W.; Ma, Y.; Meng, S.; Liu, C.; Long, L.; Tang, S. A comparative study on wear and corrosion behaviour of HVOF- and HVAF-sprayed WC–10Co–4Cr coatings. *Surf. Eng.* **2017**, *33*, 63–71. [CrossRef]

coatings

MDPI

Article

The Effect of the Laser Incidence Angle in the Surface of L-PBF Processed Parts

Sara Sendino [1,*], Marc Gardon [2], Fernando Lartategui [3], Silvia Martinez [1] and Aitzol Lamikiz [1]

1 Aerospace Advance Manufacturing Research Center-CFAA, P. Tecnológico de Bizkaia 202, 48170 Zamudio, Spain; silvia.martinez@ehu.eus (S.M.); aitzol.lamikiz@ehu.eus (A.L.)
2 EMEA AM Applications Manager, Renishaw Ibérica, Carrer de la Recerca 7, 08850 Barcelona, Spain; Marc.Gardon@renishaw.com
3 UO Structures & Statics, ITP Aero, P. Tecnológico de Bizkaia 300, 48170 Zamudio, Spain; fernando.lartategui@itpaero.com
* Correspondence: sara.sendino@ehu.eus

Received: 29 September 2020; Accepted: 22 October 2020; Published: 24 October 2020

Abstract: The manufacture of multiple parts on the same platform is a common procedure in the Laser Powder Bed Fusion (L-PBF) process. The main advantage is that the entire working volume of the machine is used and a greater number of parts are obtained, thus reducing inert gas volume, raw powder consumption, and manufacturing time. However, one of the main disadvantages of this method is the possible differences in quality and surface finish of the different parts manufactured on the same platform depending on their orientation and location, even if they are manufactured with the same process parameters and raw powder material. Throughout this study, these surface quality differences were studied, focusing on the variation of the surface roughness with the angle of incidence of the laser with respect to the platform. First, a characterization test was carried out to understand the behavior of the laser in the different areas of the platform. Then, the surface roughness, microstructure, and minimum thickness of vertical walls were analyzed in the different areas of the platform. These results were related to the angle of incidence of the laser. As it was observed, the laser is completely perpendicular only in the center of the platform, whilst at the border of the platform, due to the incidence angle, it melts an elliptical area, which affects the roughness and thickness of the manufactured part. The roughness increases from values of *Sa* = 5.489 µm in the central part of the platform to 27.473 µm at the outer borders while the thickness of the manufactured thin walls increases around 40 µm.

Keywords: roughness; incidence angle; additive manufacturing; L-PBF; INCONEL718

1. Introduction

Laser technology has been widely used in different industrial processes for many years, due to the characteristics of the laser source: high energy density, high efficiency, large temperature gradients, high repeatability of the process, and formation of a narrow heat-affected zone (HAZ) [1]. Nowadays, it is possible to find various research works on laser technologies such as laser welding [2,3], cutting [4,5], remelting [6,7], additive technologies [8,9] . . . thus demonstrating the importance of the development of these technologies.

Due to the great development that the different technologies based on metal additive manufacturing (AM) have had in recent years, it seems reasonable that one of the most researched and promising metal AM process is the Laser Powder Bed Fusion (L-PBF) technology. This technology, also known by commercial names as Selective Laser Melting (SLM) or Direct Metal Laser Sintering (DMLS), is being applied for the direct manufacturing of functional components in different sectors [10,11].

In the L-PBF process, the final part is manufactured directly using powder as raw material. This powder is homogeneously distributed by a recoater in constant thickness layers of 20 to 60 μm. Once the recoater has distributed a layer of powder, a laser beam melts the powder within an area, which has been previously specified in the CAD (Computer-Aided Design) file. Once the complete layer has been processed, the platform lowers the thickness of a layer and the process is repeated layer-by-layer until the full part is obtained [12].

This technology presents many advantages compared to conventional machining technologies such as the ability to manufacture geometrically complex parts [10,13,14]. In addition, L-PBF also presents advantages above other AM technologies as Spierings [15] showed. One of these advantages is the wide range of alloys with very different properties and characteristics that can be processed. Thus, different applications and studies can be found with high toughness Ti-TiB composites [16,17], high-density alloys such as Cu-10Sn bronze [18], high-resistant alloys such as $Al_{85}Nd_8Ni_5Co_2$ [19], or high modulus of elasticity and mechanical strength CNTs/$AlSi_{10}Mg$ [20]. Therefore, L-PBF process is of special interest in sectors such as health and aerospace due to the possibility of obtaining complex parts in a wide range of available materials [10].

However, L-PBF technology has several limitations, particularly related to the poor surface finish and the need for final post-processing of the manufactured parts [10]. The parts manufactured using L-PBF show roughness values Ra between 3.2 and 12.5 μm [13], whereas conventional technologies obtain roughness values between 1–2 μm [11].

Surface quality is a critical attribute in the aerospace and medical sectors since surface imperfections caused by roughness could impact the performance of the part. Therefore, surface roughness is a well-researched crack initiator [12], reducing the life-span of the end part due to lower tensile and fatigue strength [10]. In the medical domain, rough surfaces can accumulate more micro-organisms than a smooth surface, requiring a finishing operation in many applications in the health sector [21]. Furthermore, roughness has a significant effect on the tribological behavior of the surface in applications where the friction between surfaces is present.

Nowadays, in order to minimize the roughness and to achieve functional parts with real applicability, finishing operations are carried out. However, the finishing operation of parts manufactured using L-PBF technology is frequently very complex due to the geometrical complexity of the parts [22]. In addition, the finishing process increases considerably the manufacturing time and cost of the part [12] since most of the finishing processes involve manual operations. Moreover, sometimes it is not possible to finish some areas such as internal ducts or hollow cavities [23]. Therefore, it is necessary to reduce the roughness as much as possible in the additive manufacturing process itself before finishing the final part.

The resultant roughness can be originated by different factors. On the one hand, the process parameters during manufacturing play an important role in the resulting roughness [22,24–26]. In this sense, it is necessary to adjust parameters such as laser power, exposure time, the distance between points, and overlapping. Some previous studies state that the use of higher laser power reduces roughness, since it increases the wettability of each layer [12], while other studies show that excessive laser power can lead to spatters of excessively oxidized particles, which would also increase roughness [23]. In addition, excessive laser power will lead to higher porosity. Thus, it is necessary to find a balance between the laser power and scan speed to ensure acceptable roughness and other properties such as porosity [11,12,23]. In addition, another process parameter that influences the roughness is the thickness of the layer used [27].

In any case, even when the optimal parameters are being used, the surface roughness can be very different depending on the position of the part in the machine [28], due to the direction of the inert gas in the manufacturing chamber [29], or the angle of incidence [28] and the geometry being processed. One of the most affected areas by roughness is the downskin area. Downskin areas are those that have not had molten material in the lower layer and have been built over non-melted powder. In this case, some of these powder particles remain partially attached to the processed layer when the layer melts and re-solidifies during manufacturing [30].

There may be other effects such as partial melting of powder particles due to heat transfer when several parts are manufactured on the same platform. When different parts are manufactured very close to each other, the accumulated heat of one part may affect the adjacent, causing its surface temperature to rise, resulting in partially melt powder particles that will attach to the surface [21]. A similar effect can be observed in the case of large parts, where a high number of layers need to be processed and the generated heat into the manufacturing chamber is also higher. It should also be noted that this roughness due to the particles attached to the surface will depend on the characteristics of the powder used [31,32].

Finally, a very relevant factor is the angle of incidence of the laser with respect to the powder layer. Since the laser beam is radiated from the top part of the machine and depending on the location of the processed area, the laser beam is focused on the powder bed at a certain incidence angle depending on the area where the part is being manufactured.

The influence of this effect has not been so analyzed by the literature in comparison with other effects such as powder distribution or process parameters. The study carried out by Kleszczynski et al. [23] analyses the impact of increasing the radial distance with respect to the center of the platform, due to the laser beam incidence angle. Since the laser beam is not completely perpendicular to the platform as can be seen in Figure 1 (figure adapted from [33,34]), an increase in the roughness of the parts is obtained.

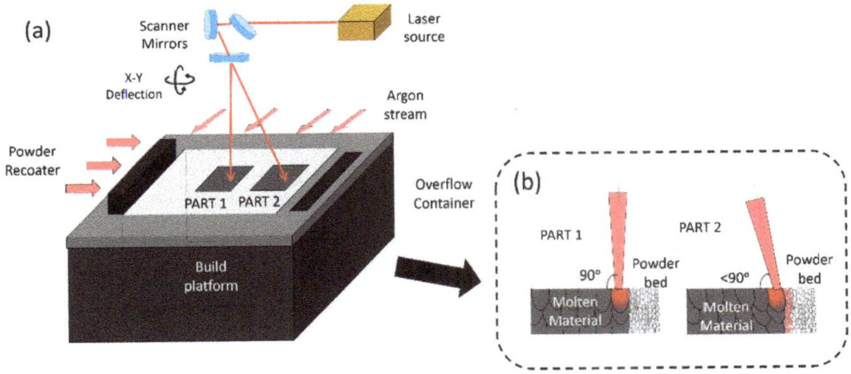

Figure 1. Effect of the incidence angle of the laser: (**a**) Diagram of the Laser Powder Bed Fusion (L-PBF) process showing the scanner operation and the possible angle of incidence in the different areas of the platform; (**b**) Detail of part 1 located in the central area of the platform and part 2 located at one side of the platform.

Throughout this article, the effect of the angle of incidence of the laser beam on the surface roughness is detailed, since, after an exhaustive bibliographic study, it seems to be a relevant but not sufficiently studied factor to minimize the resulting surface roughness obtained in the L-PBF process.

2. Materials and Methods

Two different tests were designed to analyze the incidence angle and check its effect on the final parts manufactured using L-PBF technology. All tests were carried out using the Renishaw AM400 manufacturing system (Renishaw, Stone, UK), which has a 400 W CW-M and a 3D scanner for laser beam movements with a maximum scanning speed of 7000 mm/s. This gives a reduced laser beam diameter of 70 μm in the focal point or working plane. The software used for programming the parameters of the manufacturing process was QuantAM (version V4). The tests designed for this purpose are summarized below:

2.1. First Tests: Study of the Morphology of the Laser on the Platform

This study analyzed the effect of laser radiation on the different areas of the manufacturing platform. To carry out this test, the laser beam was radiated directly to the platform, without pre-depositing a powder layer. A grid of 25 × 25 uniformly distributed points was designed in order to test the entire workspace of the platform. The parameters were adjusted to analyze the points marked on the platform separately without any overlapping, as shown in Figure 2. In addition, the correct balance between the power and exposure time parameters was set to optimize the mark on the platform. Specifically, a power of 200 W, an exposure time of 200 μs, and a distance between points of 900 μm were selected.

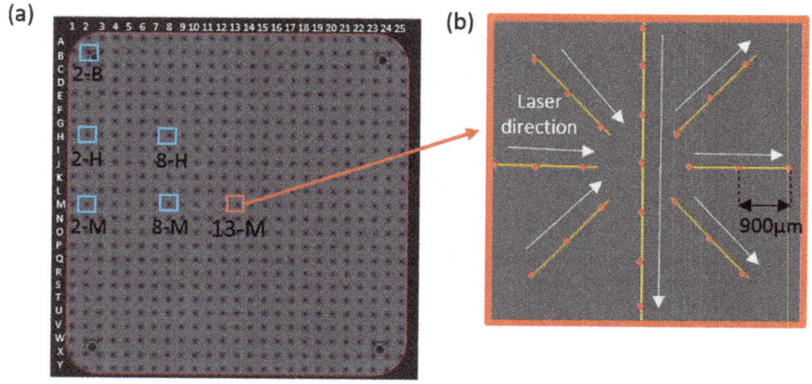

Figure 2. Test design to analyze the influence of laser inclination angle on the platform: (**a**) Distribution of the marks in the platform.; (**b**) Marks made on the platform in a star shape defining the laser strategy used.

Once the points were marked in the different areas of the platform, the morphology and dimensions of these points were studied using an infinite focus microscopy (Infinite Focus Microscope model Control Server FP G1 Vf2, Alicona, Raaba/Graz, Austria), using 500× magnification.

The different marks were made in a star shape and distributed evenly throughout the platform. With this star shape, it was intended to change the direction of the laser in order to be able to analyze the distortion caused by the angle of incidence independently of this direction of the laser and the possible delays related to the scanner.

From each of the stars marked on the platform, 10 of the points were analyzed and measured by the software Alicona MeasureSuit of the Infinite Focus Microscope measurement system. Thanks to this system, it was possible to estimate two radiuses from each of the elliptical or circular points, and by dividing these two radiuses, the aspect ratio of the ellipse could be calculated, as can be seen in Equation (1). In the case of circular marks, the value of this ratio was equal to one.

$$\text{Ellipse aspect ratio} = \frac{\text{Ellipse major axis}}{\text{Ellipse minor axis}} \tag{1}$$

Thanks to this study, it was possible to analyze the circularity of the marks made by the laser on the different parts of the platform.

2.2. Second Test: Effect of the Angle of Incidence of the Laser on Parts Manufactured by L-PBF Technology

Inconel 718 powder specially designed for this technology was used to manufacture these test parts. The powder shows high sphericity (over 80% of the particles showed 100% sphericity) and a particle size between 15–45 μm. Because of these specifications, the powder can flow and distribute correctly in each layer. In addition, powder meets the specific chemical composition of Inconel 718, shown in Table 1.

Table 1. Chemical composition in weight percentage of Inconel 718.

Elements	Ni	Cr	Co	C	Mo	Al	Ti	Fe	Nb	Si and Mn	P and S	Cu	B
%Weight	50–55	17–22	≤1	≤0.08	2.8–3.3	0.2–0.8	0.65–1.15	Bal	4.75–5.5	≤0.35	≤0.015	≤0.3	≤0.006

Three different geometries were designed and analyzed in order to study three different effects: Surface roughness, internal microstructure, and thin walls thickness.

2.2.1. Test Part A: Roughness Measurement (Red Parts)

These parts were designed specifically for measuring surface roughness, using an infinite focus 3D measurement system with 500× magnification, according to ISO 25178 [35].

The geometry designed for this purpose had a study surface of 13 mm × 13 mm, where three measurements of 1 mm × 13 mm on each of these surfaces were made, always analyzing the outside surface of each part as can be seen in Figure 3. The arithmetical mean height of the surface (Sa) and the maximum height of the surface (Sz) values were obtained and the standard deviation of these measurements was calculated to determine the error bars. In addition, thanks to this optical measurement system, surface topographies could be obtained and thus the particles partially adhered to the surface could be also analyzed.

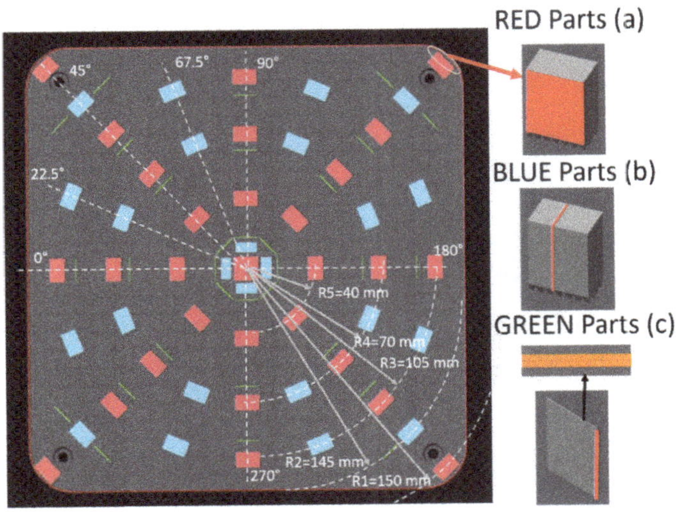

Figure 3. Design of the second platform: (**a**) parts for the roughness measurement; (**b**) parts for the microstructure analysis; (**c**) parts for measuring the thickness of thin walls.

The process parameters were the usual set for the manufacture of Inconel 718 parts using L-PBF technology using a layer thickness of 60 μm. Specifically, the volume parameters were a laser power of 200 W, exposure time of 70 μs, a distance point and hatch distance of 80 μm, a strategy angle variation of 67°, and the border parameters were a laser power of 125 W, exposure time of 75 μs, a point distance of 20 μm, and a layer thickness of 60 μm. These parts were distributed on a platform with a radial distance of 0, 40, 75, 105, and 150 mm from the center.

2.2.2. Test Part B: Microstructure Analysis (Blue Parts)

These parts were specially designed to analyze their internal microstructure. The geometry used was the same as for Parts A but the parameters were changed in order to study the desired effect. Specifically, these parts were manufactured without borders and using a 0° variation of the strategy angle so that the Gaussian-shape melted areas were aligned in order to analyze them. These parts were distributed on a platform with a radial distance of 20, 75, 105, and 145 mm from the center.

To analyze these parts, a section was cut and encapsulated. To analyze always the same surface, all parts were cut in the same direction (as shown in Figure 3b) and these samples were encapsulated using phenolic resin.

Then, the planar grinding step was made using Silicon carbide and corundum sandpaper of lower grain size of FEPA (Federation of European Producers of Abrasives) 400, to achieve a surface ready to be polished, four steps were made (using FEPA 400, FEPA 600, FEPA 800, and FEPA 1200 grain size). Once the surface was ready, it was polished using a diamond polycrystalline of 1 and 3 μm.

Once the mirror-finished surface was achieved, this surface was chemically attacked using the 25 Marble etchant using the swabbing method, as specified in the ASTM E 407 standard [36] and with a composition of 4 g $CuSO_4$, 20 cc HCl, and 20 cc H_2O. Finally, surface images were taken using the infinite focus 3D measurement system with 200× magnification.

2.2.3. Test Part C: Thin Walls Thickness Measurements (Green Parts)

Finally, thin walls of 13 mm height were designed and placed in different areas of the platform to measure their thickness. These test parts were built with a single laser track, and the same border parameters used in the "A Test Parts".

Once these parts were manufactured, the thickness of each test part was measured using the infinite focus 3D measurement system with 200× magnification. Figure 4 shows the methodology used for the analysis of thin walls. The Image J image analysis software (version ImageJ 1.x [37]) was used for extracting the contour of the images of the infinite focus microscope. This contour was then approximated to ellipses to define the thickness of the part, which would be equivalent to the average value of the vertical radius of the ellipses.

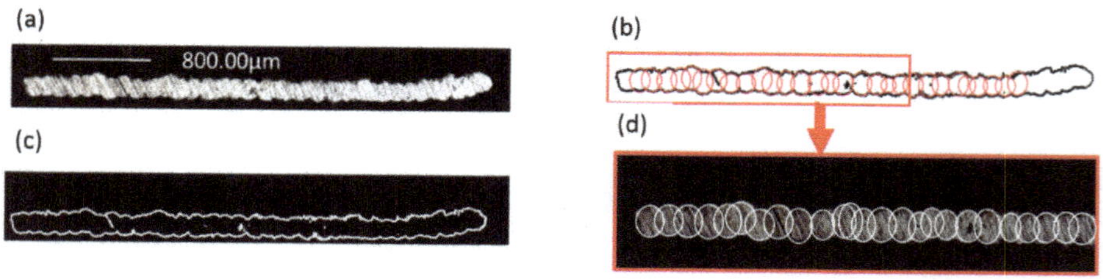

Figure 4. The methodology used for the analysis of thin walls: (**a**) image obtained through the infinite focus microscope; (**b**) extraction of the thin wall contour; (**c**) approximation of the ellipses to the generated contour; (**d**) checking this approximation in the obtained image.

Twenty-five measurements were made on each of the thin walls and the average value and standard deviation between them were calculated; therefore, and taking into account that eight thin walls were manufactured for each radial distance, 200 measurements were obtained in each of the radial distances (these radial distances being 20, 70, 100, and 145 mm).

3. Results

3.1. First Tests: Study of the Morphology of the Laser on the Platform

Analyzing the laser marks on the platform surface, it could be seen that these marks in the central part of the platform present a circular shape while marks located far from the center acquire an elliptical appearance. In addition, all the ellipses showed its largest radius in the radial direction towards the center of the platform.

Figure 5a shows this effect on the images obtained when analyzing the center point of stars with different positioning on the platform. As can be seen, the central area of the platform shows circular marks, with a diameter of around 200 µm, while in the case of the marks located away from the centre, elliptical marks up to 240 µm are observed. Therefore, a considerable difference between the different laser marks on the platform is observed depending on the location of the laser mark.

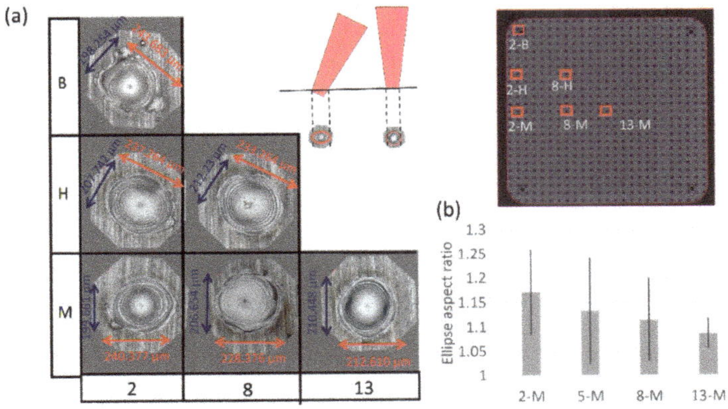

Figure 5. Analysis of the marks made by the laser on the platform: (**a**) diagram of some of the images obtained when at the central mark of the stars was analyzed (2-B, 2-H, 2-M, 8-M, 8-H, and 3-M); (**b**) ellipse aspect ratio of different stars analyzed.

In addition, the ellipse aspect ratio was calculated. It is observed in Figure 5b that this ratio was closer to unity in the tests near the center of the platform, while in the marks located on the sides of the platform, this ratio increase, and the circularity of the marks decrease.

3.2. Second Test: Effect of the Angle of Incidence on Parts Manufactured by L-PBF Technology

3.2.1. Test Part A: Roughness Measurement

The results obtained with the roughness measurement are shown in Figure 6. Despite manufacturing the same part with the same parameters and the same raw powder, surface roughness is influenced by the position of the part on the platform. As it can be observed, higher surface roughness is measured for parts located away from the center of the platform. Results show that the roughness is dependent on the position of the part on the platform because of the angle of incidence of the laser.

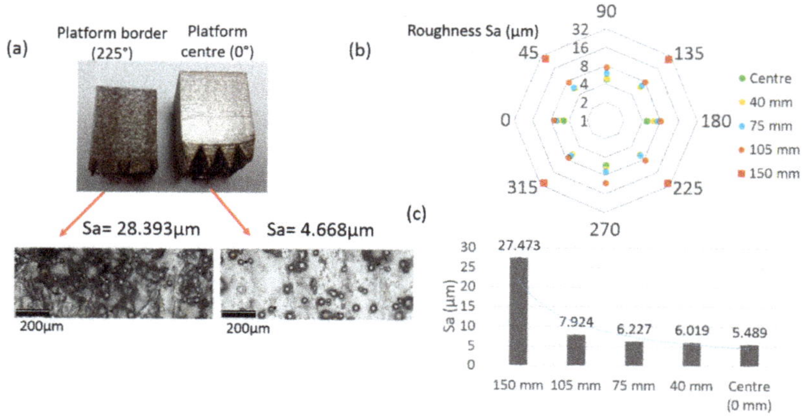

Figure 6. Roughness results obtained: (**a**) image and topographies of maximum and minimum roughness values: maximum roughness (225° and radial distance of 150 mm) and minimum roughness (center 0°); (**b**) mean value of the roughness for different radial distance from the center of the platform; (**c**) roughness mean value results obtained each test part.

As it is observed in Figure 6, the relatively high roughness and poor surface finish of the parts located at the borders of the platform can be seen with the naked eye. Analyzing the surface topographies, it could see how in the parts manufactured in the central part of the platform the layers of the part and some particles partially adhered to the surface can be differentiated; however, in the parts located in the outer border of the platform, neither the layers nor the particles partially adhered to the surface can be differentiated so clearly, but the surface quality has considerably worsened due to the melting of the powder.

It has also been determined that the roughness does not increase linearly, but increases significantly at the borders of the platform while the roughness remains at similar values in the center.

3.2.2. Test Part B: Microstructure Analysis

The microstructure has been also studied to analyze the possible influence of the incident angle on part integrity. Figure 7 shows the microstructure of two sections corresponding to the central area and the peripheral border of the platform. While the section corresponding to the part in the central area shows a uniform pattern of laser tracks, with the melted areas oriented in a vertical direction, the part in the border shows a less uniform pattern with the melted areas distorted due to the effect of the laser orientation. This effect is shown schematically in Figure 7.

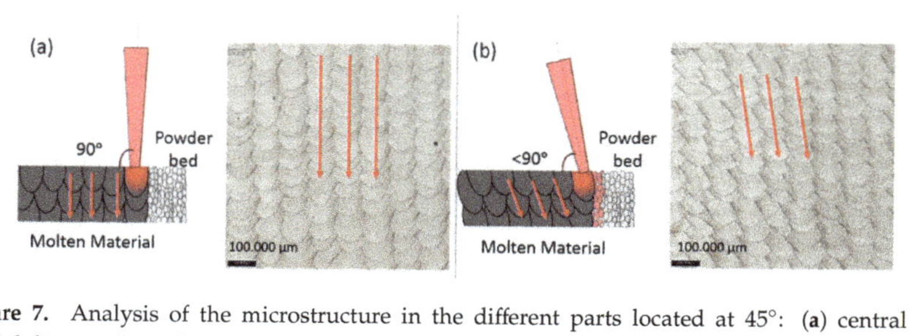

Figure 7. Analysis of the microstructure in the different parts located at 45°: (**a**) central part (radial distance from the center 20 mm); (**b**) Part located at the edge of the platform (radial distance from the center 145 mm).

Nevertheless, although the laser pattern creates a different microstructure due to the angle of the laser beam, it is observed that both components are free of porosity and defects and the mechanical properties of both specimens meet the required requirements.

3.2.3. Test Part C: Thin Walls Thickness Measurements

To conclude this study, thin walls corresponding to Test Part C and manufactured using single laser tracks were analyzed. The results of these tests are shown in Figure 8. Figure 8a shows the effect of the angle of incidence on the thickness for the different test parts manufactured in one of the diagonals of the platform (at 45°, according to Figure 3).

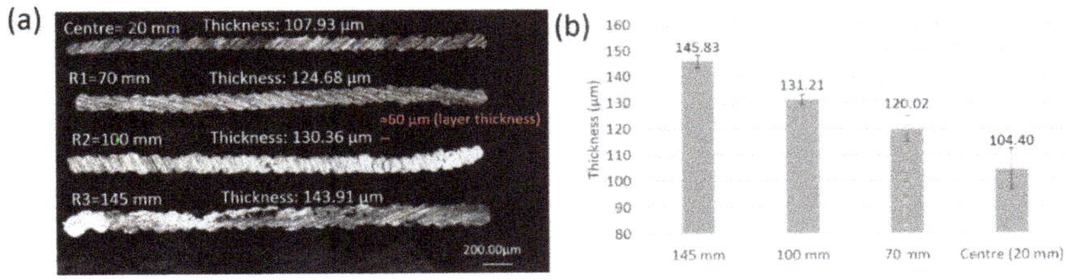

Figure 8. Analysis of the thin walls manufactured at 45°: (**a**) Measurements of wall thickness for test parts manufactured at 45° and a different distance from the platform center; (**b**) Mean value of the thickness for different radial distances from the center of the platform.

The influence of the radial distance from the center of the platform in the thickness is observed in Figure 8b) where the results show a gradual increase of the thickness as the radial distance increases. In particular, the thickness of these walls increased more than 40 µm in the parts located at the borders of the platform. Moreover, this result is consistent with the analysis of the previous tests.

4. Discussion and Conclusions

On the one hand, the measurement of the geometry of the laser spot at the platform surface demonstrated that this spot is not circular at the borders of the platform and takes an elliptical shape, which causes position-dependent laser radiation and a distortion in the melt pool. This effect is due to the technology itself, the lens used in the scanner, and the physics of the process. In this technology, and in general, in all laser technologies that require a flat working field, F-theta lenses are used, which ensure that all the main beams on the image side are parallel to the optical axis, resulting in the perpendicularity of the beam with respect to the platform.

On the other hand, results show that this angle of incidence influence the surface quality, microstructure, and thickness of the manufactured parts. As a consequence of the elliptical shape of the laser spot on the areas further away from the center, the area of powder bed heated is different from the central area and the melting of this powder does not occur in the same way. This effect is observed in Test Part A.

Furthermore, after the analysis of Test Part C corresponding to the thin walls, the effect of the angle of incidence could also be seen, as the distortion of the laser spot has resulted in a larger melted area, which has increased the thickness of the walls of the parts located in the outer borders of the platform.

Finally, after this exhaustive study, it has been possible to see how this angle of incidence has a relevant effect on the roughness of the part (Test Part B).

It should also be noted that this effect can have a considerable effect, as the increase in the diameter of the laser spot on the platform can reach up to 40 µm and different effects can occur in the manufactured part:

- Increase of the thickness of the thin walls: These walls were manufactured with a single laser track and the thickness increase was very similar to the laser spot size enlargement at the borders of the platform.
- Roughness increase: Similarly, the results of the roughness measurements are also coherent with the results of the other tests. Thus, an increase in roughness is observed as the test parts are manufactured away from the center of the platform. The high roughness, in this case, is due to the poor surface quality caused by the non-circular shape of the melt pool.
- Regarding microstructure, although a different pattern is observed due to the inclination angle and the distortion of the melt pool, similar mechanical properties are obtained, and no porosity or cracks have been observed.

Due to the great effect of the angle of incidence on the different parts manufactured and on the surface quality of the parts, it has been decided to continue with this study in future investigations. For this purpose, it is expected to characterize the laser spot on the platform, in order to determine this angle of incidence with high precision. Next, based on the shape and characteristics of the spot, the melt pool generated will be analyzed to try to obtain a direct relationship between the angle of incidence and the roughness of the parts.

Author Contributions: Conceptualization, S.S., S.M., and A.L.; methodology, S.S., S.M., M.G. and F.L.; experimental test S.S. and S.M.; result analysis, S.S., S.M., M.G., F.L. and A.L.; resources, S.M. and A.L.; writing—original draft preparation, S.S. and S.M.; writing—review and editing, S.S., S.M. and A.L.; supervision, M.G., F.L. and A.L.; project administration, M.G., F.L. and A.L.; funding acquisition, A.L. All authors have read and agreed to the published version of the manuscript.

Funding: This research was funded by the Spanish Ministry of Industry and Competitiveness under the CDTI JANO project and by Basque Government (Eusko Jaurlaritza) under the ELKARTEK Program, QUALYFAM project, Grant No. KK-2020/00042.

Conflicts of Interest: The authors declare no conflict of interest.

References

1. Kik, T. Heat source models in numerical simulations of laser welding. *Materials* **2020**, *13*, 2653. [CrossRef] [PubMed]
2. Landowski, M.; Świerczyńska, A.; Rogalski, G.; Fydrych, D. Autogenous fiber laser welding of 316L austenitic and 2304 lean duplex stainless steels. *Materials* **2020**, *13*, 2930. [CrossRef] [PubMed]
3. Sahul, M.; Tomčíková, E.; Sahul, M.; Pašák, M.; Ludrovcová, B.; Hodúlová, E. Effect of disk laser beam offset on the microstructure and mechanical properties of copper—AISI 304 stainless steel dissimilar metals joints. *Metals* **2020**, *10*, 1294. [CrossRef]
4. Wetzig, A.; Herwig, P.; Hauptmann, J.; Baumann, R.; Rauscher, P.; Schlosser, M.; Pinder, T.; Leyens, C. Fast laser cutting of thin metal. *Procedia Manuf.* **2019**, *29*, 369–374. [CrossRef]
5. Guarino, S.; Ponticelli, G.S.; Venettacci, S. Environmental assessment of selective laser melting compared with laser cutting of 316L stainless steel: A case study for flat washers' production. *CIRP J. Manuf. Sci. Technol.* **2020**. [CrossRef]
6. Temmler, A.; Pirch, N. Investigation on the mechanism of surface structure formation during laser remelting with modulated laser power on tool steel H11. *Appl. Surf. Sci.* **2020**, *526*, 146393. [CrossRef]
7. Zhang, T.; Fan, Q.; Ma, X.; Wang, W.; Wang, K.; Shen, P.; Yang, J.; Wang, L. Effect of laser remelting on microstructural evolution and mechanical properties of Ti-35Nb-2Ta-3Zr alloy. *Mater. Lett.* **2019**, *253*, 310–313. [CrossRef]
8. Gong, M.; Meng, Y.; Zhang, S.; Zhang, Y.; Zeng, X.; Gao, M. Laser-arc hybrid additive manufacturing of stainless steel with beam oscillation. *Addit. Manuf.* **2020**, *33*, 101180. [CrossRef]
9. Liu, Y.; Zhang, Y. Microstructure and mechanical properties of TA15-Ti_2AlNb bimetallic structures by laser additive manufacturing. *Mater. Sci. Eng. A* **2020**, *795*, 140019. [CrossRef]
10. Strano, G.; Hao, L.; Everson, R.M.; Evans, K.E. Surface roughness analysis, modelling and prediction in selective laser melting. *J. Mater. Process. Technol.* **2013**, *213*, 589–597. [CrossRef]

11. Li, B.-Q.; Li, Z.; Bai, P.; Liu, B.; Kuai, Z. Research on surface roughness of AlSi10Mg parts fabricated by laser powder bed fusion. *Metals* **2018**, *8*, 524. [CrossRef]
12. Mumtaz, K.; Hopkinson, N. Top surface and side roughness of Inconel 625 parts processed using selective laser melting. *Rapid Prototyp. J.* **2009**, *15*, 96–103. [CrossRef]
13. Leon, A.; Aghion, E. Effect of surface roughness on corrosion fatigue performance of AlSi10Mg alloy produced by Selective Laser Melting (SLM). *Mater. Charact.* **2017**, *131*, 188–194. [CrossRef]
14. Bosio, F.; Aversa, A.K.; Lorusso, M.; Marola, S.; Gianoglio, D.; Battezzati, L.; Fino, P.; Manfredi, D.; Lombardi, M. A time-saving and cost-effective method to process alloys by Laser Powder Bed Fusion. *J. Mater. Des.* **2019**, *181*, 107949. [CrossRef]
15. Spierings, A.B. *Powder Spreadability and Characterization of Sc- and Zr-Modified Aluminium Alloys Processed by Selective Laser Melting, Quality Management System for Additive Manufacturing, in Institute of Machine Tools and Manufacturing*; ETH Zurich: Zurich, Switzerland, 2018; p. 29. [CrossRef]
16. Attar, H.; Ehtemam-Haghighi, S.; Kent, D.; Dargusch, M.S. Recent developments and opportunities in additive manufacturing of titanium-based matrix composites: A review. *Int. J. Mach. Tools Manuf.* **2018**, *133*, 85–102. [CrossRef]
17. Dong, Y.; Li, Y.; Ebel, T.; Yan, M. Cost-affordable, high-performance Ti–TiB composite for selective laser melting additive manufacturing. *Int. J. Mater. Res.* **2020**, *35*, 1922–1935. [CrossRef]
18. Scudino, S.; Unterdörfer, C.; Prashanth, K.G.; Attar, H.; Ellendt, N.; Uhlenwinkel, V.; Eckert, J. Additive manufacturing of Cu–10Sn bronze. *Mater. Lett.* **2015**, *156*, 202–204. [CrossRef]
19. Prashanth, K.G.; Shahabi, H.S.; Attar, H.; Srivastava, V.C.; Ellendt, N.; Uhlenwinkel, V.; Eckert, J.; Scudino, S. Production of high strength $Al_{85}Nd_8Ni_5Co_2$ alloy by selective laser melting. *Addit. Manuf.* **2015**, *6*, 1–5. [CrossRef]
20. Zhao, X.; Song, B.; Fan, W.; Zhang, Y.; Shi, Y. Selective laser melting of carbon/AlSi10Mg composites: Microstructure, mechanical and electronical properties. *J. Alloys Compd.* **2016**, *665*, 271–281. [CrossRef]
21. Jamshidinia, M.; Kovacevic, R. The influence of heat accumulation on the surface roughness in powder-bed additive manufacturing. *Surf. Topogr. Metrol. Prop.* **2015**, *3*, 14003. [CrossRef]
22. Fox, J.C.; Moylan, S.P.; Lane, B. Effect of process parameters on the surface roughness of overhanging structures in laser powder bed fusion additive manufacturing. *Procedia CIRP* **2016**, *45*, 131–134. [CrossRef]
23. Jacobsmühlen, J.; Kleszczynski, S.; Ladewig, A.; Witt, G.; Merhof, D. In-situ surface roughness measurement of laser beam melted parts—A feasibility study of layer image analysis. In Proceedings of the Fraunhofer Direct Digital Manufacturing Conference (DDMC), Berlin, Germany, 16–17 March 2016; pp. 1–6.
24. Ansari, J.; Nguyen, D.-S.; Park, H.S. Investigation of SLM process in terms of temperature distribution and melting pool size: Modeling and experimental approaches. *Materials* **2019**, *12*, 1272. [CrossRef] [PubMed]
25. Yang, T.; Liu, T.; Liao, W.; Macdonald, E.; Wei, H.; Chen, X.; Jiang, L. The influence of process parameters on vertical surface roughness of the AlSi10Mg parts fabricated by selective laser melting. *J. Mater. Process. Technol.* **2019**, *266*, 26–36. [CrossRef]
26. Charles, A.; Elkaseer, A.; Thijs, L.; Hagenmeyer, V.; Scholz, S. Effect of process parameters on the generated surface roughness of down-facing surfaces in selective laser melting. *Appl. Sci.* **2019**, *9*, 1256. [CrossRef]
27. Nguyen, Q.; Luu, D.; Nai, S.; Zhu, Z.; Chen, Z.; Wei, J. The role of powder layer thickness on the quality of SLM printed parts. *Arch. Civ. Mech. Eng.* **2018**, *18*, 948–955. [CrossRef]
28. Kleszczynski, S.; Ladewig, A.; Friedberger, K.; Jacobsmühlen, J.Z.; Merhof, D.; Witt, G. Position dependency of surface roughness in parts from laser beam melting systems. In Proceedings of the 26th Internation Solid Free Form Fabrication (SFF) Symposium, Austin, TX, USA, 13–15 August 2018.
29. Ladewig, A.; Schlick, G.; Fisser, M.; Schulze, V.; Glatzel, U. Influence of the shielding gas flow on the removal of process by-products in the selective laser melting process. *Addit. Manuf.* **2016**, *10*, 1–9. [CrossRef]
30. Thijs, L.; Verhaeghe, F.; Craeghs, T.; Van Humbeeck, J.; Kruth, J.-P. A study of the microstructural evolution during selective laser melting of Ti–6Al–4V. *Acta Mater.* **2010**, *58*, 3303–3312. [CrossRef]
31. Brika, S.E.; Letenneur, M.; Dion, C.A.; Brailovski, V. Influence of particle morphology and size distribution on the powder flowability and laser powder bed fusion manufacturability of Ti–6Al–4V alloy. *Addit. Manuf.* **2020**, *31*, 100929. [CrossRef]
32. Riener, K.; Albrecht, N.; Ziegelmeier, S.; Ramakrishnan, R.; Haferkamp, L.; Spierings, A.B.; Leichtfried, G.J. Influence of particle size distribution and morphology on the properties of the powder feedstock as well as of AlSi10Mg parts produced by laser powder bed fusion (LPBF). *Addit. Manuf.* **2020**, *34*, 101286. [CrossRef]

33. Rott, S.; Ladewig, A.; Friedberger, K.; Casper, J.; Full, M.; Schleifenbaum, J.H. Surface roughness in laser powder bed fusion—Interdependency of surface orientation and laser incidence. *Addit. Manuf.* **2020**, *36*, 101437. [CrossRef]
34. Fiegl, T.; Franke, M.; Körner, C. Impact of build envelope on the properties of additive manufactured parts from AlSi10Mg. *Opt. Laser Technol.* **2019**, *111*, 51–57. [CrossRef]
35. *ISO 25178 (2016) Geometrical product specifications (GPS)—Surface Texture: Areal*; ISO Standard: Geneva, Switzerland, 2016.
36. *ASTM E407-07(2015) e1, Standard Practice for Microetching Metals and Alloys*; ASTM International: West Conshohocken, PA, USA, 2015. [CrossRef]
37. Schneider, C.A.; Rasband, W.S.; Eliceiri, K.W. NIH Image to ImageJ: 25 years of image analysis. *Nat. Methods* **2012**, *9*, 671–675. [CrossRef] [PubMed]

coatings

MDPI

Article

Microstructural and Corrosion Properties of PEO Coated Zinc-Aluminized (ZA) Steel

Luca Pezzato [1,*], Alessio Giorgio Settimi [1], Pietrogiovanni Cerchier [1], Claudio Gennari [1], Manuele Dabalà [1] and Katya Brunelli [1,2]

1 Department of Industrial Engineering, University of Padova, Via Marzolo 9, 35131 Padova, Italy; alessio.settimi@unipd.it (A.G.S.); pietro.cerchier@unipd.it (P.C.); claudio.gennari@phd.unipd.it (C.G.); manuele.dabala@unipd.it (M.D.); katya.brunelli@unipd.it (K.B.)

2 Department of Civil, Environmental and Architectural Engineering, University of Padova, Via Marzolo 9, 35131 Padova, Italy

* Correspondence: luca.pezzato@unipd.it; Tel.: +39-04-9827-5498

Received: 26 March 2020; Accepted: 28 April 2020; Published: 4 May 2020

Abstract: Plasma Electrolytic Oxidation (PEO) is a surface treatment, similar to anodizing, that produces thick oxide films on the surface of metals. In the present work, PEO coatings were obtained on zinc-aluminized (ZA) carbon steel using a solution containing sodium silicate and potassium hydroxide as electrolyte, and working with high current densities and short treatment times in Direct Current (DC) mode. The thickness of the coating, as well as the surface morphology, were strongly influenced by the process parameters, with different dissolution grades of the ZA layer depending on the current density and treatment time. A compromise between thickness and porosity of the coating was found with low current density/long treatment time or high current density/short treatment time. The PEO layer was mainly composed of aluminum oxides and silicon compounds. The corrosion resistance increased remarkably in the samples with the PEO coating. These PEO coated samples are suitable for sealing treatments that further increase their corrosion properties or will be also an ideal substrate for commercial painting, assuring improved mechanical adhesion and protection even in the presence of damages.

Keywords: plasma electrolytic oxidation; PEO; coatings; steel; zinc-aluminized; corrosion

1. Introduction

Steel is often employed in engineering applications due to its good mechanical properties, good machinability, and low cost. Carbon steels are often used in structural applications. Corrosion problems often affect carbon steels and the substitution with stainless steels is often not possible due to their higher cost. In order to overcome this problem, a possible approach is a proper surface treatment on the metal. Plasma electrolytic oxidation (PEO) of metals is an electrochemical process that produces an oxide ceramic layer on the surface. PEO is similar to traditional anodic oxidation but works with higher voltages and current densities. The high voltage (that has to be above the dielectric breakdown potential of the oxide layer) forms anodic micro-discharges that moving randomly over the processed surface produce the growth of the coating [1,2]. The corrosion and wear properties depend on current density, voltage, treatment time, and electrolyte composition employed during PEO treatment [3]. The process has been widely studied on aluminum and magnesium alloys and many results are reported in the literature, especially about the increased corrosion and wear performances of light alloys after PEO treatment. A lot of studies have been performed also on Ti, Zr, Nb, Ta and alloys obtaining very interesting results, especially in the biomaterials field [4–7]. Moreover, PEO offers the possibility to proper functionalize the surface by adding particles or specific compounds in the electrolyte [8–10]. In comparison to PEO on light alloys, the number of works regarding the application of this treatment on steels is quite low, and, in detail, some works on carbon steel and only one on low alloyed steel can be found [11–15]. These research works showed that the quality of the coatings produced directly on steels, in terms of adhesion, homogeneity, and corrosion properties, resulted lower than the ones obtained on light alloys. One of the possibilities studied in literature to overcome this problem and to produce PEO coatings with good quality on ferrous metals is to perform, before the PEO, a pre-treatment. In particular, some works can be found regarding the realization of PEO coatings on aluminized steels [16–18]. The main problem regarding this approach is that the aluminizing of steels is not such a common treatment. In order to use as a pre-treatment an economic and common treatment, zinc-based pre-treatments could be used as possible substrates for PEO coatings. However, preliminary research in the literature showed that PEO coatings formed on pure zinc have a high level of defects, as evidenced by Stojanovic et al. [19]. and are slightly protective against corrosion, as stated Rocca et al. [20]. In the literature, promising results were found for PEO coatings produced on zinc-aluminum alloys. In detail, Guangdong Bian et al. [21]. found that the PEO method can produce a continuous and dense coating on the ZA27 alloy using silicate, aluminate, and aluminate/borate electrolytes; and Guangyin Li et al. [22] showed that PEO coatings can effectively protect the ZA27 alloy from abrasive and adhesive wear and corrosion. Zinc-Aluminum coatings (ZA) are quite common in industrial applications to provide corrosion protection on steels. Considering this, the aim of this work was to study PEO process on steels with a ZA pre-treatment and to test the corrosion resistance of the obtained samples. This research represents an innovative use of PEO process, since no works in literature reported the obtainment of PEO coatings on zinc-aluminized steels. Based on the obtained results, it will be possible to obtain PEO coatings with low defect level, adherent to the substrate and characterized by a porous surface on steels, employing as pre-treatment a common treatment as ZA. The presence of pores and micro-cracks is typical of PEO coatings as extensively studied in the literature and as stated for example by Curran et al. [2]. This porosity is a characteristic of these coatings and can be technologically employed with proper post-treatments to functionalize the surface.

The PEO-coated steels, obtained in this work, could be suitable for other treatments that can further increase their corrosion properties such as sealing treatments or painting treatments. In fact, sealing treatments are commonly employed to improve corrosion protection of PEO-coated specimens [8,9]. Moreover, the presence of a PEO layer will increase mechanical adhesion of eventual commercial painting and will assure protection in case of scratches or other kind of damages, improving significantly the corrosion properties [23]. The obtained PEO coated steels can be employed, after proper post treatments, in applications where both corrosion and abrasion resistance are required.

2. Materials and Methods

The samples of carbon steel (0.16% C, 1.06% Mn, 0.09% Al, 0.03% Cr, 0.07% Cu) coated with commercial ZA coating (75% Al and 25% Zn) were used for the experiments. PEO treatments were performed using as electrolyte an aqueous alkaline solution with 20 g/L of Na_2SiO_3 and 10 g/L of KOH.

The PEO process was performed with a DC power supply of 400 V/8 A capacity (TDK-Lambda, Achem, Germany). During the treatment, the substrate was the anode and a carbon steel mesh was the cathode. The electrolyte was maintained at ambient temperature with a thermostatic bath. The treatments were performed in DC galvanostatic mode, letting the potential free to vary until a limit of 350 V, working at high current densities for short treatment times. Different current densities and different treatment times were tested. In detail, current densities of 1.1, 1.7 and 2.3 A/cm^2 were employed. For each current density, two different treatment times were used; 2 and 3 min. The samples, after the treatment, were washed with deionized water and ethanol, then dried with compressed air. The choice of the electrical parameters was performed on the base of preliminary tests and with the objective to work at high current densities and short treatment times. From the preliminary tests for high current densities and treatment times over 3 min, the coating resulted too porous. For treatment times lower than 2 min, the coating does not have the time to form (because it takes some time to start at the micro-arc stage, which is the stage where PEO coating grow). Therefore, in this work was analyzed the above reported combination of current densities and treatment times.

For all the samples, morphology, composition, thickness, and corrosion resistance of the coatings were evaluated. The samples were compared with the ones with only the ZA coating. All the samples were produced in triplicate in order to assure reproducibility.

The treated samples were cut in a cross-section and mounted in epoxy resin, then polished with standard metallographic technique (grinding with abrasive papers from 320 to 1200 grit and polishing with clothes and diamond suspension of 6 and 1 μm). Both the surface and the cross-section of treated samples were examined with a Cambridge Stereoscan 440 scanning electron microscope (Leica Microsystem S.r.l., Milan, Italy), equipped with a Philips PV9800 Energy Dispersive X-Ray Spectroscopy (EDS), (Leica Microsystem S.r.l., Milan, Italy), in order to evaluate morphology, thickness, and elemental composition of the coating. To better evaluate the distribution of the elements along the cross-section EDS elemental mapping was also performed. The phase composition of the most significant sample was performed with X-ray diffraction (XRD) using a Siemens D500 X-ray diffractometer, (Siemens, Munich, Germany) using Cu Kα radiation with a step scan of 0.05 and a counting time of 5 s in an angular range between 20° and 80°.

The corrosion performance of the coatings was determined firstly by potentiodynamic polarization tests (PDP), and then by electrochemical impedance spectroscopy (EIS) at ambient temperature.

The PDP tests were performed in a solution containing 0.1 M Na_2SO_4 with an AMEL 2549 Potentiostat (Amel Electrochemistry S.r.l., Milan, Italy). A saturated calomel electrode was employed as a reference electrode (SCE) and platinum as a counter electrode, with a scan rate of 0.5 mV s^{-1}. Considering the insulating nature of the PEO layer PDP was employed only for qualitative evaluation of the corrosion performances, whereas quantitative considerations were performed with EIS measurements [9]. The electrolyte was chosen in order to simulate atmospheric corrosion conditions in a middle aggressive environment as reported in [24].

The EIS measurements were carried out with the same cell and electrolyte employed in potentiodynamic polarization, at the value of the open circuit potential and in a frequency range between 10^5 and 10^{-2} Hz, with a perturbation amplitude of 10 mV. A Materials Instrument Spectrometer coupled with the 2549 Potentiostat was used to record EIS measurements (Amel Electrochemistry S.r.l., Milan, Italy). and the Z-View software (version 3.3) was employed for the fitting of impedance data. All the electrochemical measurements were performed in triplicate, in order to assure the reproducibility of the results.

3. Results and Discussion

3.1. Microstructural Characterization

The surfaces and the cross sections of the samples were analyzed at scanning electron microscopy (SEM) and the results are presented in Figure 1 (surfaces) and Figure 2 (cross sections).

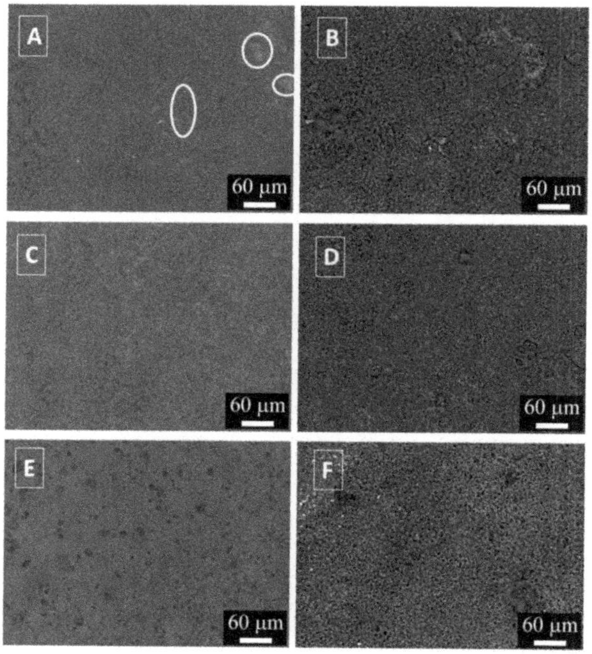

Figure 1. SEM images (backscattered electrons mode) of the surfaces of the samples: (**A**) Sample treated at 1.1 A/cm^2 for 2 min; (**B**) sample treated at 1.1 A/cm^2 for 3 min; (**C**) Sample treated at 1.7 A/cm^2 for 2 min; (**D**) Sample treated at 1.7 A/cm^2 for 3 min; (**E**) Sample treated at 2.3 A/cm^2 for 2 min; (**F**) Sample treated at 2.3 A/cm^2 for 3 min.

The surface analysis showed the typical morphology of PEO coatings with the presence of nodular structures and with the evidence of a lot of pores and micro-cracks of different dimensions, coming from the discharge phenomena. In the sample treated at 1.1 A/cm^2 for 2 min (Figure 1A) are present some uncoated zones, highlighted by white circles, indicating that the PEO layer was not completely formed on the surface. In detail, these uncoated zones are the lighter in the SEM micrographs. Increasing the treatment time up to 3 min a more uniform coating was produced, characterized by the presence of nodular structures as observable in (Figure 1B). Some pores of little dimension can be also observed. Considering the samples treated at 1.7 A/cm^2 (Figure 1C,D), in both samples the coating covered the entire surface. Comparing the two samples, an increase in the treatment time produced an increase of the nodular structures on the surface (see the right side of Figure 1D). The porosity in the sample observed in Figure 1D appear higher than the one in the sample observed in Figure 1B. The sample treated at 2.3 A/cm^2 for 2 min (Figure 1E) was characterized by a smoother surface with the presence of less but larger volcano-like pores, in comparison with the other samples. The coating seemed also more compact that the one obtained on other samples. The sample treated at 2.3 A/cm^2 for 3 min (Figure 1F) was instead characterized by the presence of many small pores and a low number of nodular structures, even if the surface was not as smooth as the one obtained at 2 min. Clearly, the sample reported in Figure 1F is the one with the highest number of pores.

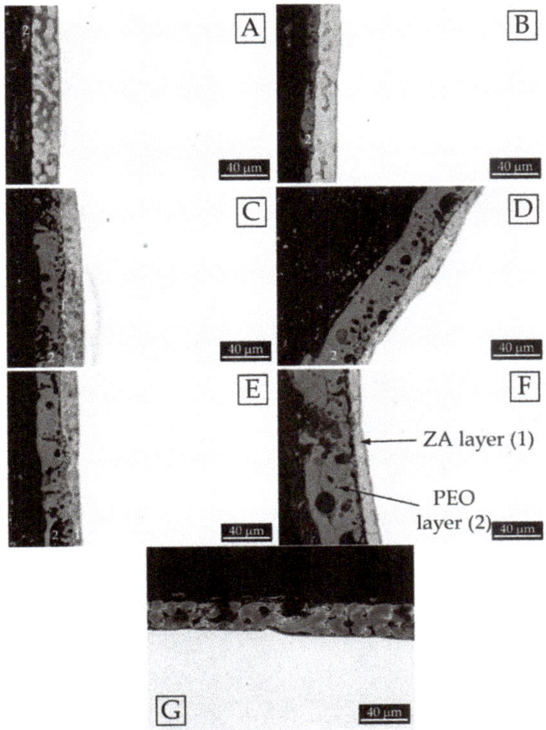

Figure 2. SEM images (backscattered electrons mode) of the cross sections of the different samples: (**A**) Sample treated at 1.1 A/cm^2 for 2 min; (**B**) Sample treated at 1.1 A/cm^2 for 3 min; (**C**) Sample treated at 1.7 A/cm^2 for 2 min; (**D**) Sample treated at 1.7 A/cm^2 for 3 min; (**E**) Sample treated at 2.3 A/cm^2 for 2 min; (**F**) Sample treated at 2.3 A/cm^2 for 3 min; (**G**) Sample with only the ZA layer In the micrographs the number (1) indicates the inner ZA layer and the number (2) the external PEO layer.

The cross sections of the samples (Figure 2), resulted in increasing current density, with a progressive thinning of the thickness of the ZA layer (the internal grey one, named 1 in the figures) and a thickening of the PEO layer (the dark and porous layer on the top, named 2). The ZA layer was about 30 µm thick in the sample without PEO layer (Figure 2G) and was about 7 µm thick in the sample treated at the higher current density and for longer treatment time (Figure 2F). This can be correlated with the oxidation of the ZA layer during PEO process and, consequently, an increase in the current density induced a higher oxidation rate. From the reported micrographs the presence of some dark grey zones in the ZA layer was observed, both in the samples before, and after, the PEO treatment. In the sample treated at 1.1 A/cm^2 for 2 min (Figure 2A) the formation of the PEO layer was not complete, whereas after the treatment for 3 min (Figure 2B) the PEO coating formed above all the surface. The samples treated at 1.7 and 2.2 A/cm^2 showed a remarkable reduction in the thickness of the ZA layer if compared with the samples obtained at 1.1 A/cm^2. Moreover, longer treatment times, at these current densities, induced the formation of thicker layers with a higher porosity in the upper layer. A summary of the thickness of the PEO layers and of the ZA layers for the various samples is reported in Table 1. A possible problem, already found in literature with PEO coatings on metallized steel [25], could be the detachment between the ZA layer and the PEO layer. In order to avoid this, a careful control of the process parameters must be performed because the problem occurs when a too thick PEO layer is present. In this case, some problems of detachment can be found in Figure 2F for the sample treated at high current density for long treatment time.

Table 1. Thickness of ZA and PEO layers of the different samples. Data coming from three different measurements.

Sample		ZA Layer Thickness (μm)	PEO Layer Thickness (μm)
Sample treated at 1.1 A/cm^2	2 min	25 ± 2	7-Not Uniform
	3 min	20 ± 2	10 ± 3
Sample treated at 1.7 A/cm^2	2 min	15 ± 1	22 ± 3
	3 min	11 ± 3	30 ± 1
Sample treated at 2.3 A/cm^2	2 min	12 ± 3	25 ± 1
	3 min	7 ± 1	40 ± 2

In order to study the composition of the different layers, EDS micro-analysis was performed on the cross section of the samples treated at the higher and the lower current density. The semi-quantitative results are reported in Table 2.

Table 2. EDS semi-quantitative results (wt %) on the various samples.

Sample and Zone			Al%	Si%	Zn%	O%
1.1 A/cm^2	2 min	ZA layer (1) light grey areas	74	-	26	-
		ZA layer (1) dark grey areas	65	-	15	20
		PEO layer (2)	60	4	16	20
	3 min	ZA layer (1) light grey areas	76	-	24	-
		ZA layer (1) dark grey areas	67	-	13	20
		PEO layer (2)	18	39	4	39
2.3 A/cm^2	2 min	ZA layer (1) light grey areas	74	-	26	
		ZA layer (1) dark grey areas	65	-	17	18
		PEO layer (2)	14	33	-	53
	3 min	ZA layer (1) light grey areas	75	-	25	-
		ZA layer (1) dark grey areas	65	-	16	19
		PEO layer (2)	18	29	2	51

Considering the results reported in Table 2, it can be noticed that the light grey areas in the inner layer were composed of only Zn and Al that is the ZA layer. The dark grey zones in the inner layer were constituted by Al, Zn and O, suggesting that oxidation phenomena occurred. In all the samples the external layer represented the actual PEO layer, with the main presence of silicon and aluminum oxides, in accordance with the composition of the electrolyte and substrate.

To further investigate the distribution of the elements in the layers, also EDS elemental mapping was performed on the sample that seemed to exhibit the lower porosity and higher uniformity of the coating, that is the one obtained at 2.3 A/cm^2 for 2 min. The results are reported in Figure 3. From the analysis of the elemental mapping, it can be observed that the PEO layer (the outer layer) was mainly composed of silicon and aluminum oxides, whereas the inner layer (the ZA layer) was composed by aluminum and zinc.

XRD analysis was also performed on the sample obtained at 2.3 A/cm^2 for 2 min. The XRD pattern is reported in Figure 4. It can be noticed the presence of the peaks of aluminum (PDF-2 database, reference N°000040787) and zinc (PDF-2 database, reference N°000040831) coming from the ZA layer, due to the penetration of X-rays under the PEO layer. From the observation of the XRD pattern, amorphous phases were present in the PEO layer, as typical for this kind of coatings [26]. Considering the crystalline part, the PEO layer resulted mainly composed by SiO_2 (PDF-2 database, reference N°000821573), in accordance with the composition of the electrolyte that was a silicate-based one. Comparing the results of EDS elemental mapping (Figure 3) and XRD (Figure 4), it is reasonable to think that the amorphous part is mainly composed by an [Al–Si–O] phase. The presence of this kind of phase, like mullite but amorphous, is also in accordance with what found by Guan et al. [27].

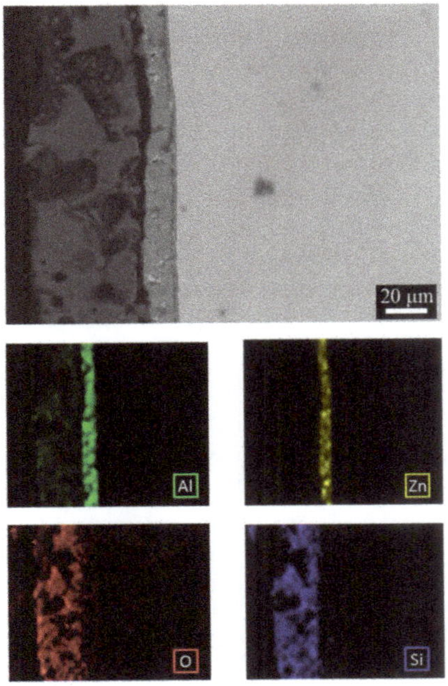

Figure 3. SEM-EDS elemental mapping performed on the cross section of the sample obtained at 2.3 A/cm^2 for 2 min.

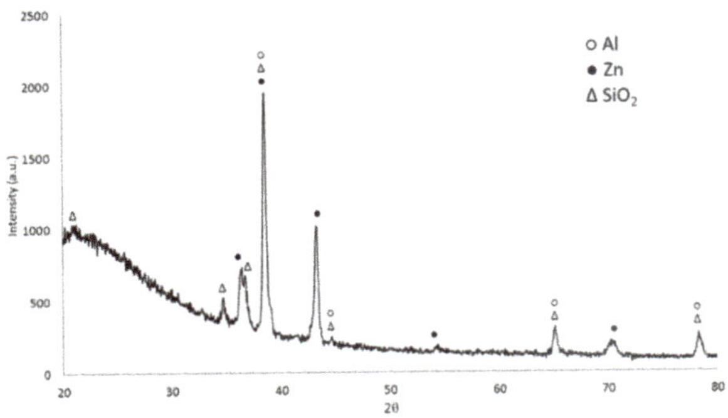

Figure 4. XRD analysis of the sample obtained at 2.3 A/cm^2 for 2 min.

Summarizing the results of the characterization of the samples, the morphology of the coatings was rich in pores, pancake, and nodular structures, in accordance with the results of other authors regarding PEO coatings produced on bulk ZA alloys [21,22]. As reported by Clyne et al. [25], most of PEO coatings present porosity, which is deleterious for the durability of the coating but, at the same time, increases stiffness and biocompatibility of the coating and, moreover, can be sealed with proper post treatments. The porosity forms because of oxygen entrapment in molten alumina in the vicinity of localized electrical discharges, which occur during PEO coating formation [2]. The discharges occurring during PEO have a strong tendency to take place in extended sequences (cascades) at fixed locations [28]. Increasing the current density generally induces a higher number of discharges location, and, consequently, of the porosity [29], and higher growth rate of the coating [30]. Also extending the treatment time produces an increase in the volcano-like pore population, at least until a critical value, as evidenced by Al Bosta et al. [31]. Treatment time also influences the thickness of the coating, as it becomes thicker with time. This is in accordance with the results reported in this work, where the increase in the current density employed in the treatment, as well as in the treatment time, produced a thinning of the ZA layer and a thickening of the PEO layer. Considering this, a compromise must be found in order to obtain a thick but not too porous coating. In detail, the samples obtained at 1.1 A/cm^2 for 3 min and the one obtained at 2.3 A/cm^2 for 2 min seemed the most promising. The PEO coating resulted composed of crystalline SiO_2 and of amorphous [Al-Si-O] phase. The presence of both the phases is in accordance with Clyne et al. [25], that showed that the composition of PEO layers is determined both by the composition of the substrate (so by the oxidation of the ZA layer in our case) and by the incorporation of compounds from the electrolyte (Silicon compounds from the silicate-based electrolyte).

3.2. Corrosion Resistance

The corrosion resistance of the samples was preliminarily evaluated by potentiodynamic polarization tests, in a solution containing sulphates. The results are reported in Figure 5. Considering that the three different curves for each sample were very similar, only one was reported for each sample.

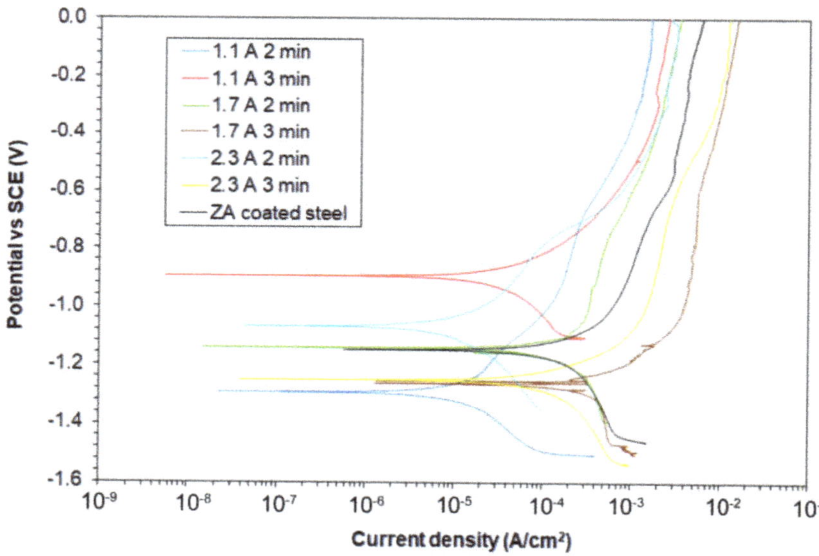

Figure 5. Potentiodynamic polarization curves for the different samples (test solution: 0.1 M Na_2SO_4).

Firstly, it is possible to see that the samples treated at 1.7 and 2.3 A/cm^2 for 3 min were characterized by a lower corrosion resistance if compared with the ZA sample, in accordance with the SEM observation that showed high porosity on these samples. The sample characterized by the higher corrosion resistance was the sample treated at 1.1 A/cm^2 for 3 min. Moreover, also the samples 1.1 A/cm^2 for 2 min and 2.3 A/cm^2 for 2 min showed better corrosion performances if compared with the ZA sample. However, considering the insulating nature of the PEO coating, no quantitative considerations can be performed from potentiodynamic polarization.

In order to deeply study the corrosion behavior of the samples also EIS tests were performed and the results in terms of Nyquist plot are reported in Figure 6. The experimental data coming from EIS were fitted with the software Z-view, using the equivalent circuits reported in Figure 7. In detail, the circuit in Figure 7A was used for the sample with only the ZA coating, with the circuit that represents the natural oxide film formed on the surface of the sample, whereas the circuit in Figure 7B was employed for the PEO treated samples. This circuit is the typical one used to fit data of PEO-treated samples, where are present an inner barrier layer, that gives the major protection against corrosion, and an external porous layer [15]. The ZA layer under the PEO coating was not considered in the equivalent circuit, since this layer, for its conductive nature, acts with the substrate as one unique electrochemical unit, under the PEO layer. Considering more in detail the meaning of the different elements in this circuit (Figure 7B): R_1 represents the resistance of the electrolyte, whereas the two parallels R_2-CPE_1 and R_3-CPE_2 consider the two different interfaces formed between the electrolyte and the two layers of the PEO coatings. The R_2-CPE_1 circuit represents the interface between the electrolyte and the external porous layer, and the R_3-CPE_2 the interface between the electrolyte into the pores and the inner barrier layer of PEO coating. As the measured capacitance is not ideal, CPE_i instead of capacitances were used in the equivalent circuits. The fitting results of the experimental data are reported in Table 3. Good fitting quality was obtained considering the low reported values of chi squared.

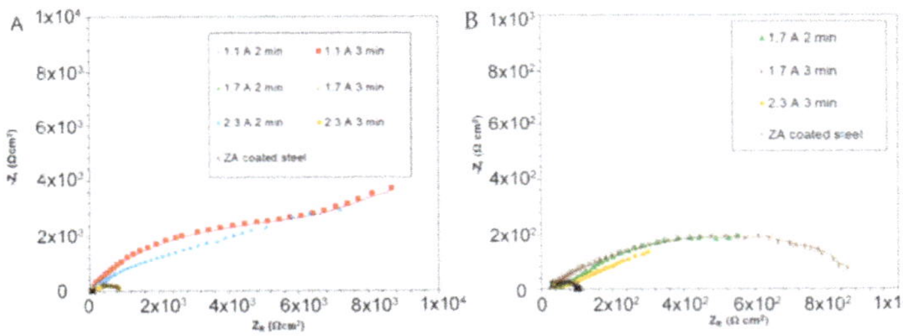

Figure 6. Nyquist plot coming from EIS tests, global results (**A**) and detail of the samples with low resistance (**B**) (test electrolyte: 0.1 M Na_2SO_4). In the graphs, dots represent the experimental data, lines the result of the fitting.

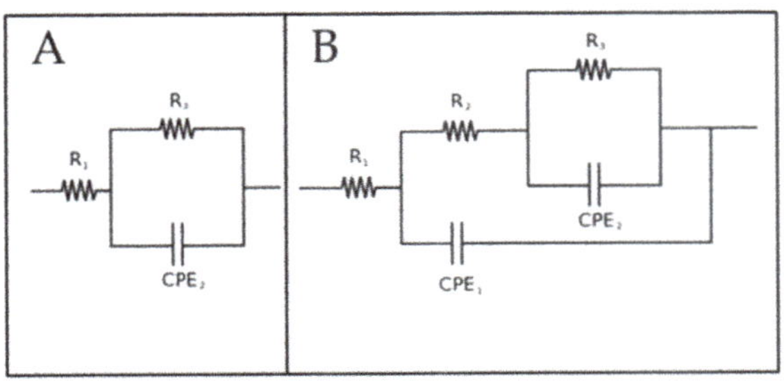

Figure 7. Equivalent circuits employed to fit EIS data for the ZA coated sample, with the circuit that represents the natural oxide film on the surface (**A**) and for the PEO treated samples, with the two circuits that represent the inner and external layers of PEO coating (**B**).

Table 3. Results of the fitting of experimental data. Standard deviation range of the single data are between 1% and 5%.

Parameter	ZA	1.1 A/cm^2		1.7 A/cm^2		2.3 A/cm^2	
		2 min	3 min	2 min	3 min	2 min	3 min
R_1 ($\Omega\cdot$cm^2)	31.2	15.1	50.2	50.5	50.2	40.2	50.4
R_2 ($\Omega\cdot$cm^2)	-	341.8	8006	51.2	15.1	1668	100.1
R_3 ($\Omega\cdot$cm^2)	77.5	1827	12,253	3241	908.4	10,241	1023
Q_1 (FHz^{1-n})	-	3.1×10^{-4}	3.6×10^{-6}	4.8×10^{-4}	3.6×10^{-6}	1.7×10^{-5}	1.5×10^{-4}
Q_2 (FHz^{1-n})	5.1×10^{-5}	5.2×10^{-4}	3.7×10^{-4}	3.8×10^{-4}	3.7×10^{-4}	9.3×10^{-5}	5.1×10^{-4}
n_1	-	0.5	0.9	0.8	0.9	0.6	0.5
n_2	0.7	0.8	0.5	0.6	0.5	0.4	0.6
Chi-squared	0.001	0.002	0.006	0.02	0.006	0.002	0.003

In analyzing the Nyquist plots, reported in Figure 6A, it resulted that the samples treated at 1.1 A/cm^2 for 3 min and the one treated at 2.3 A/cm^2 for 2 min were characterized by improved corrosion resistance in comparison with all the other samples. In fact, considering the width of the semicircle and the intersection with the X-axis as qualitative evaluation of the polarization resistance R_p, an increase of more than one order of magnitude in the polarization resistance of these two samples in comparison with the others can be observed. Observing the Nyquist plots of all the other samples (Figure 6B), an increase in the polarization resistance of all the PEO-treated samples in comparison with the only ZA one was recorded. The quality of the fitting was positive, as can be observed from the good agreement between experimental and fitting data and from the low chi-squared values.

From the analysis of the fitting results (Table 3), it was confirmed that all the PEO-treated samples were characterized by improved corrosion performances compared to the ZA sample, as evidenced by the increased values of polarization resistance (R_2 and R_3) that were one or two orders of magnitude higher in the PEO treated samples. Considering the different PEO-treated samples and comparing the values of R_2 and R_3, it can be noticed that the samples treated at 1.1 A/cm^2 for 3 min and treated at 2.3 A/cm^2 for 2 min showed values of R_3 one order of magnitude higher than the other PEO-treated samples. This can be correlated with the higher homogeneity and lower porosity of these samples, as resulted by SEM observations. In particular, large differences among the various samples were found in the R_3 value and these can be linked with the porosity: R_3 represents the resistance of the inner barrier layer of PEO coatings, when the porosity is low (samples treated at 1.1 A/cm^2 for 3 min and treated at 2.3 A/cm^2 for 2 min) only small quantity of electrolyte can penetrate and attack this layer. When instead a high number of pores is recorded also the inner layer is clearly more prone to corrosion, and the R_3 value strongly decreases.

In general, thickness and porosity are the main factors that can influence the corrosion properties of PEO coated samples. The increase in coating thickness played an important role in the corrosion behavior due to the barrier effect given by the coating. However, high current density, applied for long treatment times resulted detrimental for corrosion protection due to the coarse and very porous microstructure, as already found in literature on another substrate such as ZK60 alloy [32]. Also, Bala Srinivasan et al. [33] found that high current density increased the number of defects and micro-cracks and that a compromise between growth rate and quality of the microstructure must be found. The results of the present work agreed with these considerations. In fact, it was found that the better results in terms of corrosion resistance were obtained either with low current density and long treatment time (sample treated at 1.1 A/cm^2 for 3 min) or with high current density and short treatment time (sample treated at 2.3 A/cm^2 for 2 min) that were the samples that exhibited good compromise in term of thickness and porosity.

4. Conclusions

The present work demonstrated the possibility of producing PEO coatings on zinc-aluminized steels. PEO coatings produced on pure zinc, from the literature, were characterized by a high level of defects and are not protective against corrosion. The acceptable homogeneity and low defect level, obtained in this work, have to be ascribed to the presence of aluminum in the ZA layer, which dissolves and builds up the PEO structure. The better conditions in order to maximize the corrosion resistance resulted 1.1 A/cm^2 for 3 min and 2.3 A/cm^2 for 2 min. In this way, a homogeneous coating, mainly composed of amorphous [Al–Si–O] phase and crystalline silicon oxide, was formed on the top of the ZA layer. The improved corrosion performances of these two samples, in comparison with the others, can be ascribed to the good compromise between coating thickness and porosity. Considering that ZA treatment is a common treatment on carbon steel, the use of ZA layer as pre-treatment layer may represent a promising way to produce PEO coating of good quality on steels. Eventual problems could be the adhesion between the ZA layer and the PEO layer, and the high economic cost of PEO process.

Author Contributions: Conceptualization, L.P. and P.C.; methodology, L.P.; software, A.G.S.; validation, L.P. and A.G.S.; formal analysis, L.P. and A.G.S.; investigation, L.P. and A.G.S.; resources, M.D. and K.B.; data curation, A.G.S. and C.G.; writing—original draft preparation, L.P. and A.G.S.; writing—review and editing, L.P., K.B.; visualization, C.G.; supervision, M.D. and K.B.; project administration, M.D. and K.B.; funding acquisition, M.D. and K.B. All authors have read and agreed to the published version of the manuscript.

Funding: This research received no external funding.

Conflicts of Interest: The authors declare no conflict of interest.

References

1. Martin, J.; Melhem, A.; Shchedrina, I.; Duchanoy, T.; Nominé, A.; Henrion, G.; Czerwiec, T.; Belmonte, T. Effects of electrical parameters on plasma electrolytic oxidation of aluminium. *Surf. Coat. Technol.* **2013**, *221*, 70–76. [CrossRef]
2. Curran, J.A.; Clyne, T.W. Porosity in plasma electrolytic oxide coatings. *Acta Mater.* **2006**, *54*, 1985–1993. [CrossRef]
3. Pezzato, L.; Coelho, L.B.; Bertolini, R.; Settimi, A.G.; Brunelli, K.; Olivier, M.; Dabalà, M. Corrosion and mechanical properties of plasma electrolytic oxidation-coated AZ80 magnesium alloy. *Mater. Corros.* **2019**, *70*, 2103–2112. [CrossRef]
4. Fattah-Alhosseini, A.; Keshavarz, M.K.; Molaei, M.; Gashti, S.O. Plasma Electrolytic Oxidation (PEO) Process on Commercially Pure Ti Surface: Effects of Electrolyte on the Microstructure and Corrosion Behavior of Coatings. *Met. Mater. Trans. A* **2018**, *49*, 4966–4979. [CrossRef]
5. Sowa, M.; Simka, W. Effect of DC Plasma Electrolytic Oxidation on Surface Characteristics and Corrosion Resistance of Zirconium. *Materials* **2018**, *11*, 723. [CrossRef]
6. Sowa, M.; Kazek-Kęsik, A.; Krząkała, A.; Socha, R.; Dercz, G.; Michalska, J.; Simka, W. Modification of niobium surfaces using plasma electrolytic oxidation in silicate solutions. *J. Solid State Electrochem.* **2018**, *18*, 3129–3142. [CrossRef]
7. Sowa, M.; Simka, W. Electrochemical Impedance and Polarization Corrosion Studies of Tantalum Surface Modified by DC Plasma Electrolytic Oxidation. *Materials* **2018**, *11*, 545. [CrossRef]
8. Mingo, B.; Arrabal, R.; Mohedano, M.; Llamazares, Y.; Matykina, E.; Yerokhin, A.; Pardo, A. Influence of sealing post-treatments on the corrosion resistance of PEO coated AZ91 magnesium alloy. *Appl. Surf. Sci.* **2018**, *433*, 653–667. [CrossRef]
9. Pezzato, L.; Rigon, M.; Martucci, A.; Brunelli, K.; Dabalà, M. Plasma Electrolytic Oxidation (PEO) as pre-treatment for sol-gel coating on aluminum and magnesium alloys. *Surf. Coatings Technol.* **2019**, *366*, 114–123. [CrossRef]
10. Pezzato, L.; Cerchier, P.; Brunelli, K.; Bartolozzi, A.; Bertani, R.; Dabalà, M. Plasma electrolytic oxidation coatings with fungicidal properties. *Surf. Eng.* **2019**, *35*, 325–333. [CrossRef]
11. Saikiran, A.; Hariprasad, S.; Arun, S.; Rama Krishna, L.; Rameshbabu, N. Effect of electrolyte composition on morphology and corrosion resistance of plasma electrolytic oxidation coatings on aluminized steel. *Surf. Coatings Technol.* **2019**, *372*, 239–251.
12. Wang, Y.; Jiang, Z.; Yao, Z.; Tang, H. Microstructure and corrosion resistance of ceramic coating on carbon steel prepared by plasma electrolytic oxidation. *Surf. Coatings Technol.* **2010**, *204*, 1685–1688. [CrossRef]

13. Wang, Y.; Jiang, Z.; Yao, Z. Preparation and properties of ceramic coating on Q235 carbon steel by plasma electrolytic oxidation. *Curr. Appl. Phys.* **2009**, *9*, 1067–1071. [CrossRef]
14. Wang, Y.; Jiang, Z.; Yao, Z. Microstructure, bonding strength and thermal shock resistance of ceramic coatings on steels prepared by plasma electrolytic oxidation. *Appl. Surf. Sci.* **2009**, *253*, 650–656. [CrossRef]
15. Pezzato, L.; Brunelli, K.; Dolcet, P.; Dabalà, M. Plasma electrolytic oxidation coating produced on 39NiCrMo3 steel. *Surf. Coatings Technol.* **2016**, *307*, 73–80. [CrossRef]
16. Wu, Z.; Xia, Y.; Li, G.; Xu, F. Structure and mechanical properties of ceramic coatings fabricated by plasma electrolytic oxidation on aluminized steel. *Appl. Surf. Sci.* **2007**, *253*, 8398–8403. [CrossRef]
17. Karpushenkov, S.A.; Shchukin, G.L.; Belanovich, A.L.; Savenko, V.P.; Kulak, A.I. Plasma electrolytic ceramic-like aluminum oxide coatings on iron. *J. Appl. Electrochem.* **2010**, *40*, 365–374. [CrossRef]
18. Gu, W.C.; Lv, G.H.; Chen, H.; Chen, G.L.; Feng, W.R.; Zhang, G.L.; Yang, S.Z. Preparation of ceramic coatings on inner surface of steel tubes using a combined technique of hot-dipping and plasma electrolytic oxidation. *J. Alloys Compd.* **2007**, *430*, 308–312. [CrossRef]
19. Stojadinović, S.; Tadić, N.; Vasilić, R. Formation and characterization of ZnO films on zinc substrate by plasma electrolytic oxidation. *Surf. Coatings Technol.* **2016**, *307*, 650–657. [CrossRef]
20. Rocca, E.; Veys-Renaux, D.; Guessoum, K. Electrochemical behavior of zinc in KOH media at high voltage: Micro-arc oxidation of zinc. *J. Electroanal. Chem.* **2015**, *754*, 125–132. [CrossRef]
21. Bian, G.; Wang, L.; Wu, J.; Zheng, J.; Sun, H.; DaCosta, H. Effects of electrolytes on the growth behavior, microstructure and tribological properties of plasma electrolytic oxidation coatings on a ZA27 alloy. *Surf. Coatings Technol.* **2015**, *277*, 251–257. [CrossRef]
22. Li, G.; Mao, Y.; Li, Z.; Wang, L.; DaCosta, H. Tribological and Corrosion Properties of Coatings Produced by Plasma Electrolytic Oxidation on the ZA27 Alloy. *J. Mater. Eng. Perform.* **2018**, *27*, 2298–2305. [CrossRef]
23. Cerchier, P.; Pezzato, L.; Gennari, C.; Moschin, E.; Moro, I.; Dabalà, M. PEO coating containing copper: A promising anticorrosive and antifouling coating for seawater application of AA 7075. *Surf. Coat. Technol.* **2020**, *393*, 125774. [CrossRef]
24. Saha, J.K. *Corrosion of Constructional Steels in Marine and Industrial Environment*; Springer: New Delhi, India, 2013; pp. 129–143; ISBN 978-81-322-0720-7.
25. Clyne, T.W.; Troughton, S.C. A review of recent work on discharge characteristics during plasma electrolytic oxidation of various metals. *Int. Mater. Rev.* **2019**, *64*, 127–162. [CrossRef]
26. Blawert, C.; Bala Srinivasan, P. Plasma Electrolytic Oxidation Treatment of Magnesium Alloys. In *Surface Engineering of Light Alloys*; Dong, H., Ed.; Woodhead Publishing Limited: Cambridge, UK, 2010.
27. Guan, Y.; Xia, Y. Amorphus coatings deposited on aluminum alloy by plasma electrolytic oxidation. *Trans. Nonferrous Met. Soc. China* **2005**, *15*, 565–570.
28. Troughton, S.C.; Nominé, A.; Dean, J.; Clyne, T.W. Effect of individual discharge cascades on the microstructure of plasma electrolytic oxidation coatings. *Appl. Surf. Sci.* **2016**, *389*, 260–269. [CrossRef]
29. Erfanifar, E.; Aliofkhazraei, M.; Nabavi, H.F.; Sharifi, H.; Rouhaghdam, A.S. Growth kinetics and morphology of plasma electrolytic oxidation coating on aluminum. *Mater. Chem. Phys.* **2017**, *185*, 162–175. [CrossRef]
30. Hussein, R.O.; Nie, X.; Northwood, D.O. An investigation of ceramic coating growth mechanisms in plasma electrolytic oxidation (PEO) processing. *Electrochim. Acta* **2013**, *112*, 111–119. [CrossRef]
31. Al Bosta, M.M.S.; Ma, K.-J.; Chien, H.-H. The effect of MAO processing time on surface properties and low temperature infrared emissivity of ceramic coating on aluminium 6061 alloy. *Infrared Phys. Technol.* **2013**, *60*, 323–334. [CrossRef]
32. Wu, H.L.; Cheng, Y.L.; Li, L.L.; Chen, Z.H.; Wang, H.M.; Zhang, Z. The anodization of ZK60 magnesium alloy in alkaline solution containing silicate and the corrosion properties of the anodized films. *Appl. Surf. Sci.* **2007**, *253*, 9387–9394. [CrossRef]
33. Bala Srinivasan, P.; Liang, J.; Blawert, C.; Störmer, M.; Dietzel, W. Effect of current density on the microstructure and corrosion behaviour of plasma electrolytic oxidation treated AM50 magnesium alloy. *Appl. Surf. Sci.* **2009**, *255*, 4212–4218. [CrossRef]

coatings

MDPI

Article

Fabrication of Zinc Substrate Encapsulated by Fluoropolyurethane and Its Drag-Reduction Enhancement by Chemical Etching

Yuanzhe Li [1], Zhe Cui [2], Qiucheng Zhu [3], Srikanth Narasimalu [4] and Zhili Dong [1,*]

1 School of Materials Science & Engineering, Nanyang Technological University, Singapore 639798, Singapore; yuanzhe001@e.ntu.edu.sg
2 Department of Materials Science and Engineering, Zhengzhou University, Zhengzhou 450001, China; cuizhe@zzu.edu.cn
3 School of Chemical Engineering, Sichuan University, Chengdu 610065, China; bob-lee@163.com
4 Energy Research Institute @ NTU (ERI@N), CleanTech One, Singapore 637141, Singapore; NSRIKANTH@ntu.edu.sg
* Correspondence: zldong@ntu.edu.sg; Tel.: +65-6790-6727

Received: 23 March 2020; Accepted: 9 April 2020; Published: 10 April 2020

Abstract: A fluoropolyurethane-encapsulated process was designed to rapidly fabricate low-flow resistance surfaces on the zinc substrate. For the further enhancement of the drag-reduction effect, Cu^{2+}-assisted chemical etching was introduced during the fabrication process, and its surface morphology, wettability, and flow-resistance properties in a microchannel were also studied. It is indicated that the zinc substrate with a micro-nanoscale roughness obtained by Cu^{2+}-assisted nitric acid etching was superhydrophilic. However, after the etched zinc substrate is encapsulated with fluoropolyurethane, the superhydrophobic wettability can be obtained with a contact angle of 154.8° ± 2.5° and a rolling angle of less than 10°. As this newly fabricated surface was placed into a non-standard design microchannel, it was found that with the increase of Reynolds number, the drag-reduction rate of the superhydrophobic surface remained basically unchanged at 4.0% compared with the original zinc substrate. Furthermore, the prepared superhydrophobic surfaces exhibited outstanding reliability in most liquids.

Keywords: fluoropolyurethane; zinc substrate; Cu^{2+}-assisted etching; superhydrophobic/hydrophilic; drag reduction

---

1. Introduction

Industrial pipeline drag reduction technology has been in existence for a long time, and the modern superhydrophobic interface undoubtedly provides a new research direction for drag reduction. Therefore, the efficient preparation of superhydrophobic surfaces may generally include (1) constructing suitable micro/nanostructures on low surface energy materials or (2) modifying surfaces with low surface energy materials [1].

Polyurethane has the advantages of wear resistance, tear resistance, flex resistance, etc. over many other coating materials, but it also has disadvantages, because the bubbles are easily trapped in the polyurethane due to its long curing process and high surface energy. The copolymerization of silicone and polyurethane can address these defects; e.g., using polydimethylsiloxane (PDMS) reacted with isocyanates or prepolymer formed a silicone–polyurethane copolymer through addition polymerization and a chain-extension reaction [2]. Besides, the advanced siloxane-modified polyurethane with polyamino functional groups in the side chain was also used to further enhance its original performance [3–5]. The amino siloxane attached as the side chain of the polyurethane could be conducive to the migration of silicon atoms to the surface and effectively improve the surface and mechanical properties of the polyurethane [5,6]. Although the surface energy of polysiloxane is already very small, perfluoroalkane could be even half of the polysiloxane. Therefore, the introduction of perfluoroalkyl groups into the chain of polysiloxane could not only make the surface energy as low as fluorine, but the flexibility and softness could also become similar to that of silicone [7–9].

Except for the modification of surface materials, the surface microstructure could also provide an important methodology for the surface modification. The microstructure of lotus leaves and shark-skin surfaces have been analyzed in recent years. Studies also have shown that these microstructure surfaces indicate good superhydrophobic and self-cleaning effects [10]. In 1997, German botanists Barthlott and Neinhuis [11] referred to this characteristic of the lotus leaf as the Lotus Effect. Cottin et al. [12] found that the superhydrophobic surface of this micro/nanostructure can significantly reduce the flow resistance as well. Many researchers have developed superhydrophobic surfaces by analyzing the surfaces of plant leaves, whose slipping phenomenon could significantly reduce the flow resistance. What is more, various experimental approaches are capable of creating such a microstructure interface, which include template methods, plasma texturing, lithography [13,14], and chemical deposition at present. However, the preparation process of the template method, which requires pre-preparation of the template, is so complicated that has not been widely used yet [15,16]. The processing of plasma texturing with thermal oxidation method takes so much time, and it is applicable for limited substrate materials. Although the plasma-induced nanotexture is uniform and indicates outstanding superhydrophobic wettability in the recent work of K. Ellinas et al., the good attachment of plasma fluorocarbon deposition could be obtained in polymer substrate only [17]. It is still easy to be scratched, which needed to be further enhanced with more than one polymeric layer on the commonly used metal substrates. As for the chemical deposition method, which has advantages such as its simple process and equipment, it is merely able to be applied on those metals that are less active than others [18–20]. The strong acid and alkali solution are mainly used for the chemical etching for some commercial formations of surface textures. However, the etching stability of the reaction process is poor because of the variation of the concentration, the surface microscopic size, and morphology are difficult to control [21,22].

In previous attempts of fabricating piping materials and preparing marine coatings, it is found that the modification of the surface material and deposition of a coating with a low surface energy at the original surface material can also significantly reduce the flow resistance. The lateral adhesion force between a liquid drop and a solid can also be divided into static and kinetic regimes. With the decrement of the surface energy and increment of the superhydrophobicity, the lateral adhesion force between the liquid and the surface would be reduced, and the drops would start sliding over solid surface easily [23]. Besides, the spatial surface structures formed by chemical etching and the substitution reaction would also enhance the drag-reduction effect. As the roughness increased, the threshold force to initiate the drop motion would decrease [24]. Hence, the combination of a chemical etching method and a proper coating material could not only extend the original mechanical and chemical durability by the protective coating, but its surface structure advantages could also be reserved [25]. Moreover, the drag-reduction performance of the substrate could be further enhanced. The durability of the surface, especially the chemical durability probed by immersion in liquids, is also another factor to consider regarding the fabrication. Polyvinylidene fluoride (PVDF), polytetrafluoroethylene (PTFE), and polydimethylsiloxane (PDMS) are the typical materials that are being used in the fabrication of low surface energy coating, while the superhydrophobic fluoropolyurethane surface and its drag-reduction behavior have been seldom documented [26–29].

This paper proposed a fast method to fabricate a drag-reduction fluoropolyurethane surface. The nanostructure of this drag-reduction surface could be further enhanced by a chemical etching process [27]. Through the flow-resistance test in the microchannel, such surface exhibited a drag-reduction rate of 4.0% and an outstanding reliability of surface wettability. The effects of how the surface characteristics have influenced the microchannel flow resistance are also discussed in this paper.

2. Experiment and Methods

2.1. Chemicals

Methyl perfluorooctadecanoate (PC5139M, $CF_3{\cdot}(CF_2)_{16}{\cdot}COOCH_3$) (3B Scientific Corporation, Wuhan, China), aminopropyl polydimethyl siloxane (APS, $H_2N(CH_2)_3$ $(Me_2SiO)_8{\cdot}Me_2Si(CH_2)_3NH_2$) (GE, Tokyo, Japan), polypropylene glycol (PPG, $M_n = 2000 \pm 100$, China National Pharmaceutical Group Chemicals Co., Chengdu, China), 2,4-toluene diisocyanate (TDI) (Shanghai Chemicals Co., Shanghai, China), xylene (Tianjin Chemicals Co., Tianjin, China), 3,3′-dichloro-4, 4′-Diamino-diphenylmethane (MOCA) (Japan Mitsui Chemicals Co., Tokyo, Japan), copper nitrate, anhydrous ethanol, nitric acid (Kelong Chemicals Co., Chengdu, China), and zinc substrate (1.8 mm thickness, purity ≥ 94%) (Guangzhou Yudi Scientific Instrument Co., Ltd., Guangzhou, China).

2.2. Fabrication of Drag-Reduction Substrate

2.2.1. Etching Treatment of Zinc Substrate

Firstly, the 30 mm × 30 mm zinc substrate was rinsed with acetone 5 times and ultrasonically cleaned in deionized water for 15 min. After being dried in the oven at 100 °C for 30 min, the samples were etched by a mixing solution with 5.0 mL of nitric acid, 30.0 mL of distilled water, and 1.0 g of copper nitrate for 60 s at 60.0 °C. Finally, these samples were rinsed thoroughly with deionized water and dried at 100 °C for another 30 min.

2.2.2. Fluoro-Modified Polyurethane Encapsulated Treatment

Firstly, Methyl Fluoro-Aminopropyl Polydimethyl Siloxane (MF-APS) was synthesized by Methyl perfluorooctadecanoate (PC5139M) and aminopropyl polydimethyl siloxane (APS) droplets under 50 °C and stirred at 6.7 s^{-1} for 2.5 h. After the vacuum distillation of the previous substituted process, the PPG were treated together with MF-APS by vacuum dehydration at 25 °C for another 10 min (Figure S1). Then, a proper amount of xylene and excess 2,4-toluene diisocyanate (TDI) were added into the dehydrated mixture, preheated, and reacted for another 2.5 h at 90 °C to synthesize the isocyanate-terminated polyurethane prepolymer (Figure 1) [28]. After the synthesis of polyurethane prepolymer, the melted 3,3′-dichloro-4,4′-Diamino-diphenylmethane (MOCA) was added in and stirred at 13.3 s^{-1} for 60 s. Finally, the reacted fluropolyurethane was sprayed on the etched/unetched zinc substrate, cured at 60 °C for 7 days, and then dried at room temperature for another day to prepare fluoropolyurethane treated drag-reduction surface material. Furthermore, the etched zinc substrate capped with fluoropolyurethane and unetched zinc substrate capped with and w/o fluoropolyurethane were the test surfaces, while the unetched zinc substrate served as the control group in this experiment.

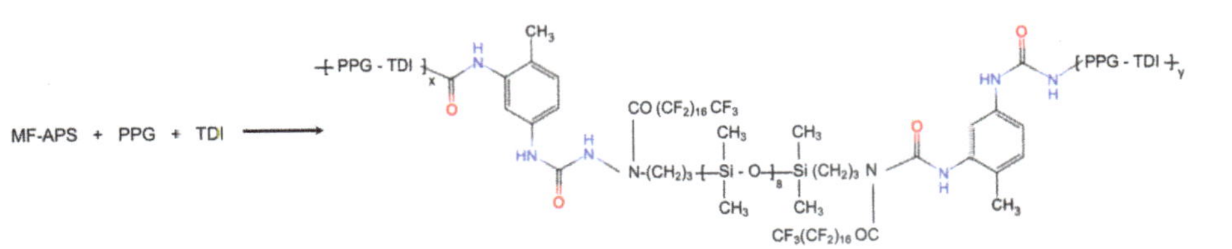

Figure 1. Synthesis of prepolymer end capped by isocyanate (–NCO) groups.

2.3. Surface Characterization

Fourier Transform Infrared Spectroscopy (FTIR spectra) was obtained at room temperature on a Nicolet-6700 spectrophotometer (Thermo Electron Corporation, Waltham, MA, USA) between 4000 and 600 cm^{-1} with the resolution of 4 cm^{-1}. The surface morphology of the samples was characterized by S-4700 SEM (Hitachi, CA, USA) and surface roughness data were obtained by an Alpha-Step D-500 stylus profiler (KLA-Tencor Corporation, Milpitas, CA, USA) with a force of 10.0 mg, speed at 0.10 mm/sec, and data points filter at 8 points. The contact angles (CAs) and surface energy (SE) were measured and analyzed by OCA15 plus and attached SCA 20 software (Data physics, Filderstadt, Germany). At the same time, continuous shooting was used to observe the spreading and rolling of water droplets on the sample surface. For the testing of wettability stability, all samples were immersed in deionized water; 8.0 wt % hydrochloric acid (acid), 8.0 wt % sodium hydroxide (alkaline), and 40.0 wt % sodium chloride (saline) solution respectively were measured for continuous 10 days [26,29–34].

2.4. Microchannel Drag Reduction Test

As the reference and unified standard for the microchannel drag reduction test was seldom documented, the microchannel test of superhydrophilic/hydrophobic surfaces was finally carried out using a non-standard design microchannel flow resistance test device. The parameters of its microchannel are shown in Figure 2. After stabilization by a regulator tank, the deionized water was pressurized with nitrogen gas and driven into the microchannel. The height of the microchannel was adjusted by tightening the screw. At each height of the microchannel, the pressure difference (ΔP) over the increment of the Reynold number was obtained. Besides, the flow resistance and drag reduction effect of the superhydrophilic/hydrophobic surface was evaluated by the pressure drop ΔP at the same Reynolds number in comparison with the unetched zinc substrate [35–38]. As the length (L) of the microchannel was a thousand times its height (h) and the Reynold number regions inside the microchannel could be varied from 50 to 1200 in this experiment, the experimental parameters for the flow-resistance test could still cross the laminar (less than 600) and turbulent phases (above 800) [39–43], and the drag reduction effect inside the microchannel could be well reflected. The uncertainty/error and parameters of the non-standard design microchannel are indicated in Table S1. Only the stable and unchanged reading would be recorded for each variation of microchannel height as well as Reynolds number.

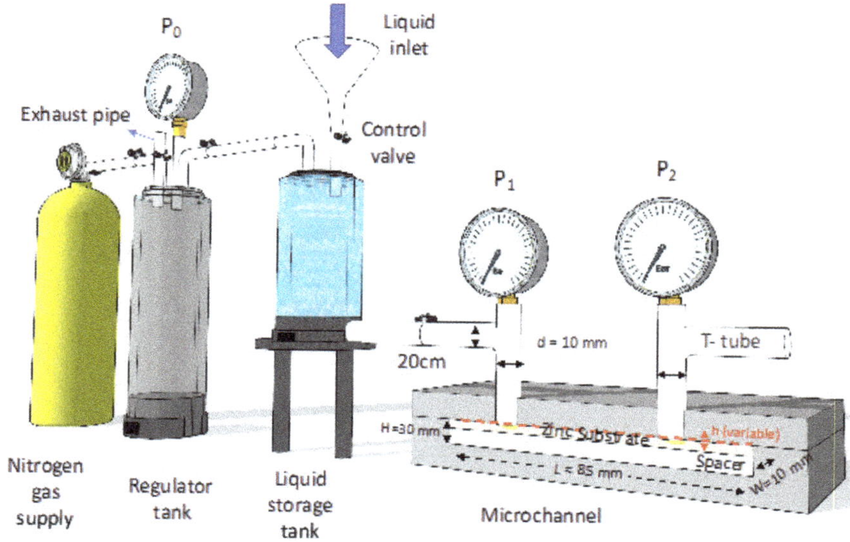

Figure 2. Non-standard design microchannel flow resistance test device.

3. Results and Discussion

3.1. FTIR Characterization of Fluoropolyurethane Prepolymer

To confirm that the fluoropolyurethane has been successfully fabricated, the FTIR spectra of the polyurethane sprayed on the etched zinc substrate was obtained (Figure 3). The MF-APS and PPG treated with TDI exhibited the absorption peaks of –NHCO between 1892 and 2232 cm^{-1}, indicating the immobilization of PPG and MF-APS in the polyurethane. The FTIR spectra of the fluoropolyurethane displayed that the typical characteristic peaks of –N=C=O, –CF_2 (C–F) and Si–O–Si function groups from the original compound appeared at 2429 cm^{-1}, 1496 cm^{-1} and 1064–1087 cm^{-1} respectively.

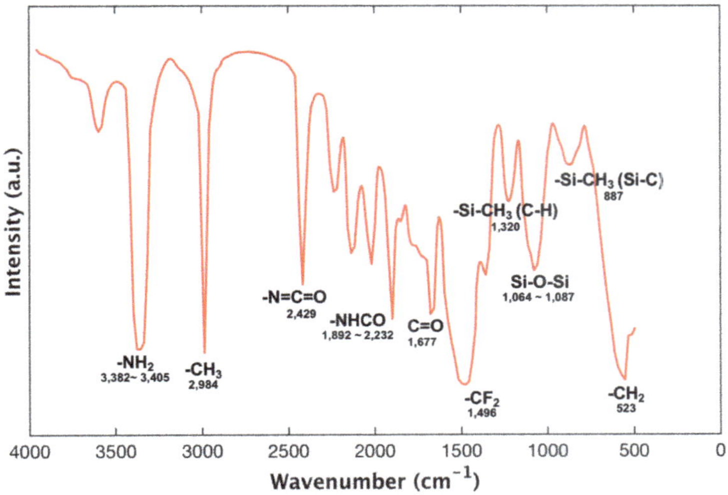

Figure 3. FTIR spectra of fluoropolyurethane.

3.2. Surface Morphology

In the case of amplifying 1000 and 50,000 times, the surface profiling of all four groups were indicated in Figure 4, the surface of the zinc substrate etched by Cu^{2+}/ HNO_3 was rough and had many convex structures, and the fluoropolyurethane demonstrates well-adhesion to zinc substrates as it could still maintain its original surface textures, as shown in the comparison between Figure 4a,b and Figure 4c,d. The trapped air bubbles in the coating because of its long curing time were addressed through the temperature control, as shown in Figure 4b,d. Moreover, in previous experiments, another group of zinc substrate structure etched by nitric acid with copper ions for 10 and 60 s had also been observed. The surface formed by corrosion for 10 s only indicated shallow texture, but the surface texture for the ones with 60 s corrosion was more prominent and deeper, which was considered as the optimal group in the trial drag-reduction tests.

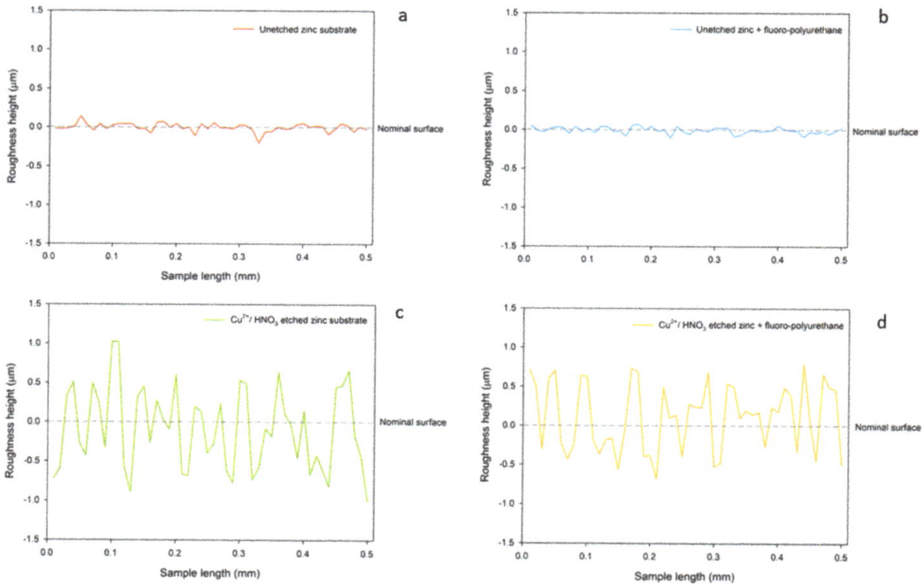

Figure 4. Surface profiling of (**a**) unetched zinc substrate, (**b**) unetched zinc + fluoropolyurethane, (**c**) Cu^{2+}/ HNO_3-etched zinc, and (**d**) Cu^{2+}/ HNO_3-etched zinc + fluoropolyurethane.

An unetched zinc substrate encapsulated with fluoropolyurethane from the Scanning Electron Micrograph (SEM) of zinc substrate had a smooth surface without too much surface structure. A comparative observation of the zinc substrate structure without the corrosion of copper ionized nitric acid solution and with copper ionized nitric acid solution is indicated in Figure 5a,h. In the case of a magnification of 50,000 times, the surface structures of the different zinc substrates were compared, and it was found that the topography of the etched ones was rough. The additional fluoropolyurethane, as shown in Figure 5c,d and Figure 5g,h, indicated a minor morphology difference compared with the Cu^{2+}/ HNO_3-etched only zinc substrate in Figure 5a,b and Figure 5e,f, and outstanding compactness on the etched surface. From the following contact angle (CA) analysis, this surface structure also contributed to the construction of superhydrophobic surfaces [23,24].

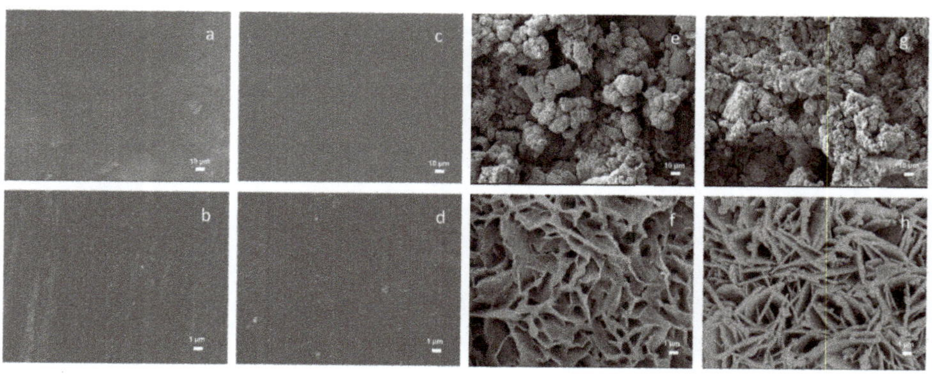

Figure 5. Scanning Electron Micrograph (SEM) of surface morphology: (**a**) Unetched zinc (×1000 times), (**b**) Unetched zinc (×50,000 times), (**c**) Unetched zinc + fluoropolyurethane (×1000 times), (**d**) Unetched zinc + fluoropolyurethane (×50,000 times), (**e**) Cu^{2+}/ HNO_3-etched zinc (×1000 times), (**f**) Cu^{2+}/ HNO_3-etched zinc (×50,000 times), (**g**) Cu^{2+}/ HNO_3-etched zinc + fluoropolyurethane (×1000 times), and (**h**) Cu^{2+}/ HNO_3 etched zinc + fluoropolyurethane (×50,000 times).

3.3. Surface Wettability Analysis

The contact angle of the zinc substrate only without any corrosion was 75.0° ± 1.8°. After being corroded by Cu^{2+}/HNO_3, the water droplets were quickly spread out, as shown in Figure 6, showing superhydrophilicity, and the contact angle was less than 3.0°.

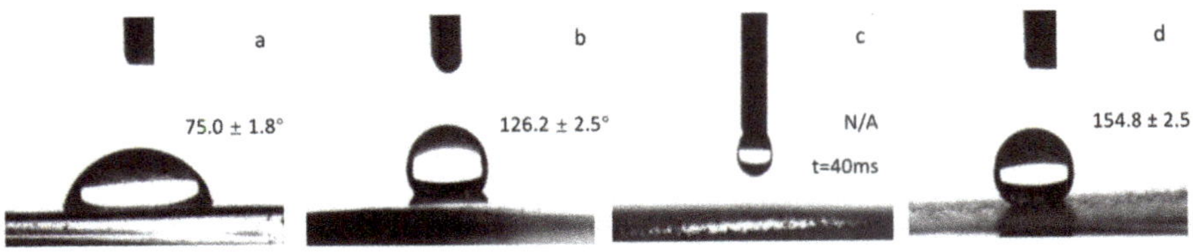

Figure 6. Surface contact angle of zinc substrate (**a**) Unetched zinc substrate, (**b**) Unetched zinc + fluoropolyurethane, (**c**) Cu^{2+}/HNO_3-etched zinc, and (**d**) Cu^{2+}/HNO_3-etched zinc + fluoropolyurethane.

As shown in Figure 6, after the fluoropolyurethane was applied, the water contact angle of the zinc substrate etched by the Cu^{2+}/HNO_3 solution and encapsulated with fluoropolyurethane could reach as high as 154.8° ± 2.5°, showing superhydrophobic properties, while water drops spread rapidly on the surface of superhydrophilic zinc substrates etched by the Cu^{2+}/HNO_3 solution only.

It was shown that a nano/micrometer superhydrophobic surface could be formed on the surface of the zinc substrate through chemical etching. Such a surface could provide good properties of superhydrophobicity and drag reduction [24]. This surface was not stained with water, and the rolling was extremely likely to occur. The zinc substrate was slightly inclined at the angle of 8.0°, and the water droplets could roll freely (Figure S2). The surface energy (SFE) result of an etched zinc substrate by diluted nitric acid with Cu^{2+} ion encapsulated with fluoropolyurethane also indicated low readings at 22.4 mJ/m^2 only, while the one without etching by any diluted nitric acid with fluoropolyurethane was at 30.5 mJ/m^2.

In the experiment of chemical durability probed by immersion in different liquids [26,34], this fluoropolyurethane coating platform was capable of maintaining a relatively long-term stability in different environments as shown in Figure 7, since immersion in liquids usually results in degradation of the surface/coating properties. It exhibited remarkable resistance to water and acid/alkaline/salt chemicals and extreme durability against immersion over 10 continuous days. Static contact angle measurements of water were performed [26,29]. Hydrophilicity is mostly due to the material characteristics and micro-nano roughness of the surface [30–33]. As long as the surface structure is not changed, it will not affect the superhydrophilicity of surface itself at all. Even comparing with the other superhydrophobic materials, the contact angle of this fluoropolyurethane coating was doing as well as polyvinylidene fluoride (PVDF), polytetrafluoroethylene (PTFE), or polydimethylsiloxane (PDMS), and it would remain at 154.8° for most of time [26,30–34]. In addition, it also indicated good adhesion on the etched zinc surface without any sign of peeling.

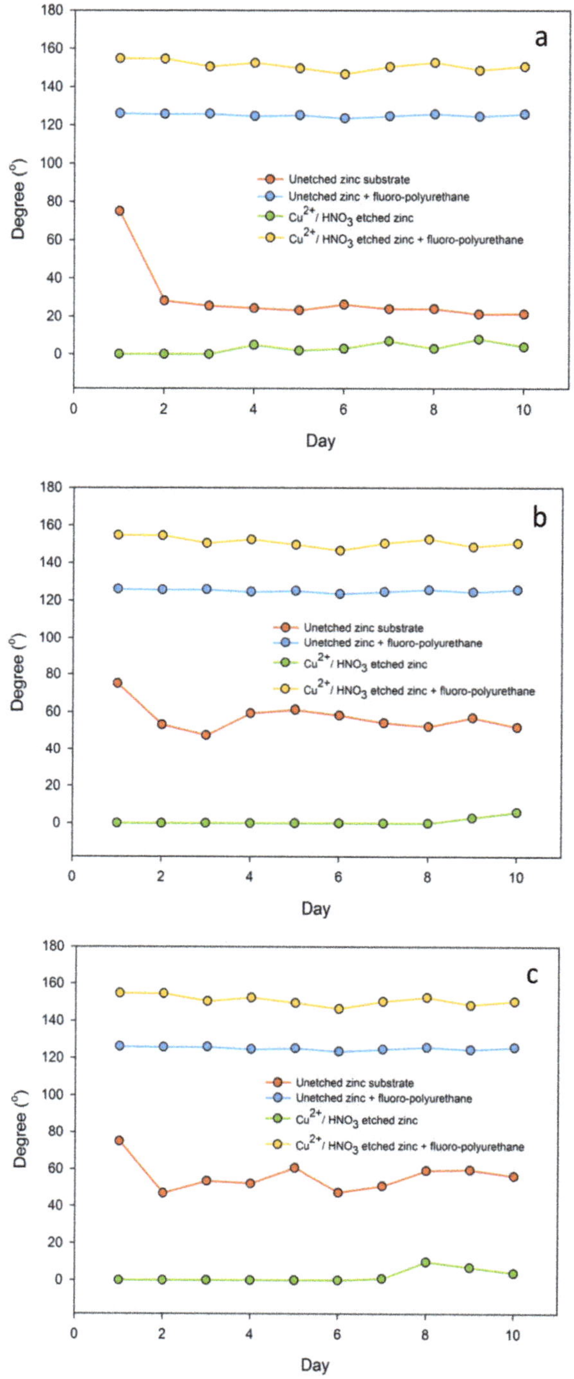

Figure 7. Surface wettability variability for continuous 10 days in three different chemicals (**a**) 8.0 wt % hydrochloric acid (acid), (**b**) 8.0 wt % sodium hydroxide (alkaline), and (**c**) 40.0 wt % sodium chloride (saline) solution.

3.4. Microchannel Drag-Reduction Analysis

The microchannel drag-reduction experiment was aimed to investigate the drag reduction effect at all the flow rates to stimulate the practical application during commercial usage. In the experiment, since it is not a circular pipe, the equivalent diameter $W' = 2Wh/(W+h)$ was used, where $W = 10$ mm and variable h were the width and height of the microchannel. In order to observe the actual flow-resistance effect, the pressure drop must be tested at the same Reynolds number. The Reynolds number (Re) is a dimensionless number that can be used to characterize fluid flow, which is expressed as Re, $Re = \varrho v d/\mu$, where v, ϱ, μ are the flow velocity, density, and viscosity coefficient of the fluid, respectively. As d was supposed to be the feature length of the microchannel, so the W' was also used as the 'feature length' (d) in calculating Re. The use of Reynolds numbers to distinguish fluid flow was laminar or turbulent and can also be used to determine the resistance to flow of an object in a fluid.

Firstly, we measure under the condition of water inlet pressure by taking the readings of the cylinder and pressure meters on the microchannel. The flow rate under inlet pressures was measured, and the pressure–flow linear relationship was demonstrated according to the formula:

$$v_z = \frac{1}{12}\frac{\Delta P h^3 W}{\mu L} \quad (1)$$

i.e., v_z was obtained through the flow volume meter, ΔP was the pressure difference between the inlet and outlet water supply, $W' = 10 \times 10^{-3}\ h\ /\ (10 \times 10^{-3} + h)$ mm, $L = 85$ mm, and μ_{water} was taken as 1.00 mPa·s. The microchannel height $h = 192.53$ μm was determined by solving the above equations [35–40].

Comparing to the relationship of the pressure drop according the increment of Reynolds number, the diagram as shown in Figure 8 was finally generated. The trendline indicated the laminar flow regions under the height of 192.53 μm, which should be in between $Re = 0$ and $Re = 580$. As the Reynolds number at the horizontal ordinate was constant, the less the pressure dropped, the less flow resistance the surface should have. Therefore, it could be observed that the pressure drops of these fluoropolyurethane-capped zinc substrates were smaller than those of the unetched zinc substrate at the same Reynolds number, which indicted a better drag reduction effect. Moreover, as the Reynolds number increased under the same channel height, the drag reduction effect would increase slightly as well.

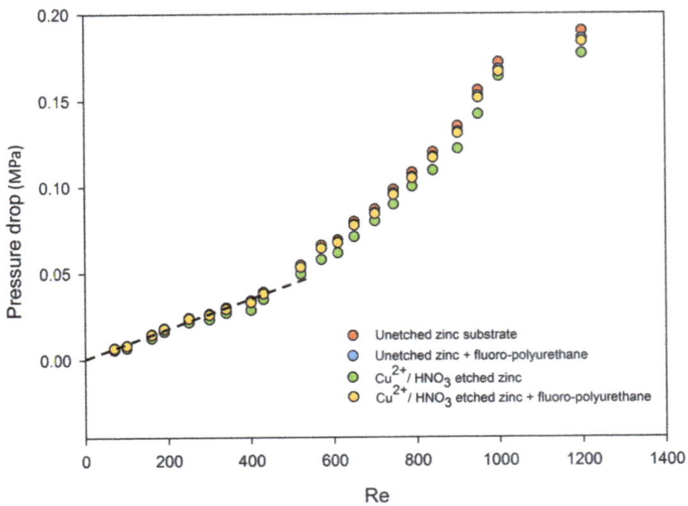

Figure 8. Pressure drop–Reynolds number diagram at $h = 192.53$ μm microchannel height.

Then, comparing with the unetched zinc substrate (control group), the drag reduction rate of the superhydrophilic and superhydrophobic zinc substrate were made. The drag reduction rate (%) was calculated using following equation:

$$\text{Drag reduction rate } (\%) = \frac{\Delta P_b - \Delta P_a}{\Delta P_b} \quad (2)$$

where ΔP_b was the pressure drop with no surface modification or corrosion (control group), ΔP_a was the pressure drop after corrosion or capped with fluoropolyurethane [41–43]. Diagrams of the drag reduction rate (%) over Reynolds number, for a total of 21 points, were plotted for each zinc substrate group. It could be observed that the superhydrophilic drag-reduction rate was decreased with the increment of the Reynolds number. The superhydrophobic drag reduction rate in the all flow zone was basically maintained at about 3.16%, as shown in Figure 9.

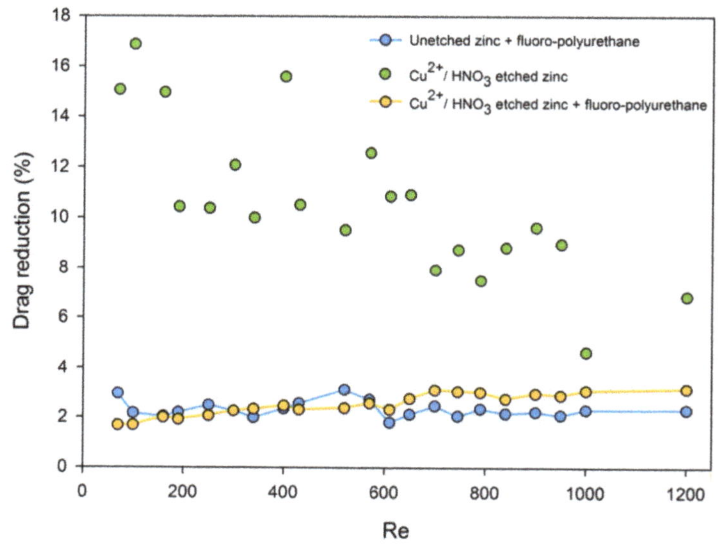

Figure 9. Drag-reduction rate–Reynolds number diagram of the superhydrophilic/hydrophobic surface compared with an unetched zinc substrate at h = 192.53 μm.

After each Reynolds-number group under h = 192.53 μm drag-reduction test was finished, the height of the channel was adjusted to reduce or increase by tightening the screws at h = 122.21 μm and h = 277.48 μm, respectively. Measurements were taken to obtain the pressure drop and Reynolds number data of different zinc substrates. Comparing the Reynolds number and pressure drop diagrams tested under different channel heights, two sets of pressure drop Reynolds number relationship diagrams were generated also, and all the data were under the same pattern [44].

It is known from Figure 10 that the resistances of all the surfaces were changed by varying the microchannel height. Under all the flow regions, from laminar to turbulent, both the superhydrophilic and superhydrophobic surfaces indicated certain drag-reduction effects. With the decrement of the flow rate or the increment of channel height, the superhydrophilic drag reduction rate became smaller, and the superhydrophobic drag reduction rate remained unchanged at around 4.0%, which was relatively stable. As for the superhydrophobic surface fabricated in this paper, it is generally believed that there is a non-shear air/water interface on the low Van der Waals force, and the fluid can shear freely at the gas–liquid interface with almost no resistance, which results in wall slippage, a lower friction coefficient, and flow resistance [23,41–43]. Since air was also able to be dissolved in the water, the non-shear air–water interface was formed by the separated air and the coating surface during the time that water flew over. Hence, for those etched surfaces, water molecules could not directly contact the surface of the fluoropolyurethane, and these surface characteristics could reduce the friction factor. As for continuously increasing the Reynolds number and flow rate, the non-shear air–water interface might become thinner and thinner [45,46].

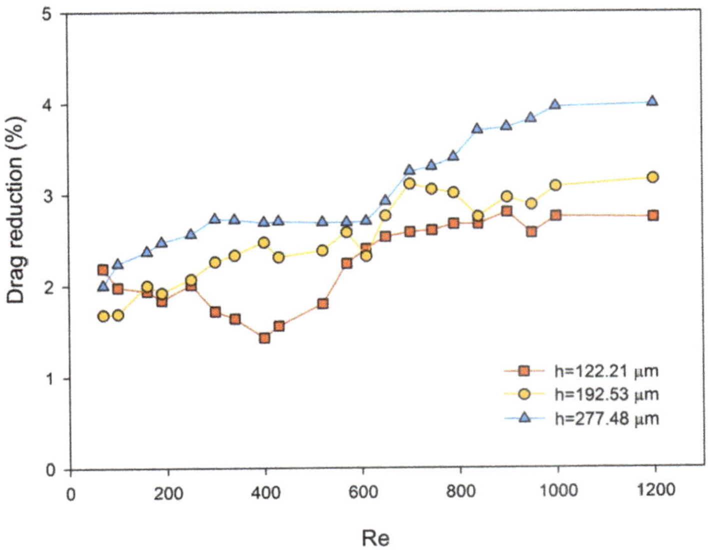

Figure 10. Cu^{2+}/HNO_3-etched zinc + fluoropolyurethane drag reduction rate at different channel heights and Reynolds numbers.

However, the superhydrophilic drag reduction effect was even more obvious than the superhydrophobic. This interesting and repeatable scenario was supposededly caused by the ripple effect regarding the dissolved gas with higher affinity and electrical polarization sticking to the superhydrophilic surface than the superhydrophobic one, an achievement of microbubble drag reduction by the structure change of the boundary layer [45–48], and surface abrasion that resulted from high fluid velocity and enabled the substituted copper particles rolling near the wall surface. This causality is still under further analysis through experimental and modeling methodology separately in ongoing work [26,43,46–48].

To conclude, the main achievement of this research regards the aspects of the novel modification of the polyurethane-coating system, the good combination between an inorganic substrate and organic coating for drag reduction, the new methodology for a flow resistance test in the microchannel, and inclusive discussion about the drag-reduction effect and mechanisms with the latest research findings. As it was known to date, superhydrophobic, hierarchical, nanotextured polymerics were applied to the surface to enhance its superhydrophobicity and drag-reduction effect [17,48]. However, the adhesion force between the organic-coating matrix and the public use metal surface is still low for most polymeric coatings. When the pressure impact to the surface becomes higher, the surface texture would be certainly destroyed, and the coating layer would be even peeled off [30–34]. In this study, the polyurethane-based coating system, which had good adhesion with zinc substrates, was fluoro-modified and then applied in the microchannel for the flow resistance test. The microchannel was designed with variable height and extremely small ratios between the length, width, and height. The experimental result and mechanism analysis indicated that at low Reynolds number flow conditions, the surface texture friction and intermolecular forces between water and fluoropolyurethane would be the main factors that influenced the drag reduction, while at high Reynolds numbers, the low surface energy and the slight non-shear air–water interface would dominate the drag-reduction effect.

4. Conclusions

This paper proposed a fast method to fabricate a superhydrophobic and low flow-resistance fluoropolyurethane surface. Its drag-reduction effect could be further enhanced by the micro-nanoscale structures on the zinc substrate, which was formed by the copper-ion assisted nitric acid etching on the zinc substrate via a simple process. As expected, the micro-nanostructures obtained by chemical etching could contribute to the enhancement of the superhydrophobicity of the surface characters and low flow resistance.

As this newly fabricated surface was applied into a non-standard design microchannel, the drag reduction rate remained basically unchanged at 4.0% with the all variations of Reynolds numbers. Furthermore, the prepared superhydrophobic surface exhibited outstanding surface durability when it was immersed in solutions of acid, alkali, and salt.

Supplementary Materials: The following are available online at http://www.mdpi.com/2079-6412/10/4/377/s1, Figure S1: Synthesis of prepolymer and fluoro-polyurethane, Figure S2. Waterdrops rolled on Cu2+/HNO3-etched zinc + fluoropolyurethane (rolling angle 8 ± 1.6°), Table S1. Uncertainty/error and parameters of non-standard design microchannel.

Author Contributions: Conceptualization, Y.L.; Data curation, Y.L.; Formal analysis, Y.L.; Funding acquisition, S.N. and Z.D.; Investigation, Y.L.; Methodology, Q.Z.; Resources, Y.L. and Q.Z.; Supervision, Z.C., S.N. and Z.D.; Writing – original draft, Y.L.; Writing – review & editing, Y.L., Z.C., S.N. and Z.D. All authors have read and agreed to the published version of the manuscript.

Funding: This research was funded by MOE Academic Research Fund (AcRF) Tier 1 Project "Nano-structured Titania with tunable hydrophilic/hydrophobic behavior and photocatalytic function for marine structure application", Grant Call (Call 1/2018) _MSE (EP Code EP5P, Project ID 122018-T1-001-077), Ministry of Education (MOE), Singapore.

Conflicts of Interest: The authors declare no conflict of interest.

References

1. Ma, M.; Randal, M. Superhydrophobic surfaces. *Curr. Opin. Colloid Interface Sci.* **2006**, *11*, 193–202. [CrossRef]
2. Ong, C.; Murthe, S.S.; Mohamed, N.; Perumal, V.; Mohamed Saheed, M. Nanoscaled Surface Modification of Poly(dimethylsiloxane) Using Carbon Nanotubes for Enhanced Oil and Organic Solvent Absorption. *ACS Omega* **2018**, *3*, 15907–15915. [CrossRef] [PubMed]
3. Ge, Z.; Luo, Y. Synthesis and characterization of siloxane-modified two-component waterborne polyurethane. *Prog. Org. Coat.* **2013**, *76*, 1522–1526. [CrossRef]
4. Philipp, C. Waterborne polyurethane wood coatings based on rapeseed fatty acid methyl esters. *Prog. Org. Coat.* **2012**, *74*, 705–711. [CrossRef]
5. Chattopadhyay, D.K.; Raju, K.V.S.N. Structural engineering of polyurethane coatings for high performance applications. *Prog. Polym. Sci.* **2007**, *32*, 352–418. [CrossRef]
6. Mathew, A.; Kurmvanshi, S.; Mohanty, S.; Nayak, S. Preparation and characterization of siloxane modified: Epoxy terminated polyurethane-silver nanocomposites. *Polym. Compos.* **2018**, *39*, E2390–E2396. [CrossRef]
7. Galhenage, T.; Webster, D.; Moreira, A.; Burgett, R.; Stafslien, S.; Vanderwal, L.; Clare, A. Poly(ethylene) glycol-modified, amphiphilic, siloxane–polyurethane coatings and their performance as fouling-release surfaces. *J. Coat. Technol. Res.* **2017**, *14*, 307–322. [CrossRef]
8. Zhou, X.X.; Sun, B.; Wu, S.; Zhang, X.; Liu, Q.; Xiao, Y. Reports on Materials Science Findings from Wuhan University of Technology Provide New Insights (Evaluation On Self-healing Mechanism and Hydrophobic Performance of Asphalt Modified By Siloxane and Polyurethane).Report. *J. Wuhan Univ. Technol. Mater. Sci. Ed.* **2019**. [CrossRef]
9. Noreen, A.; Zia, K.; Tabasum, S.; Aftab, W.; Shahid, M.; Zuber, M. Structural elucidation and biological aptitude of modified hydroxyethylcellulose-polydimethyl siloxane based polyurethanes. *Int. J. Biol. Macromol.* **2020**, *150*, 426–440. [CrossRef]

10. Feng, L.; Li, S.; Li, Y. Super-hydrophobic surfaces from natural to artificial. *Adv. Mater.* **2002**, *14*, 1857–1860 [CrossRef]
11. Barthlott, W.; Neinhuis, C. Purity of the sacred lotus, or escape from contamination in biological surfaces *Planta* **1997**, *202*, 1–8. [CrossRef]
12. Cottin, B.; Barrat, J.; Bocquet, L. Low-friction flows of liquid at nanopatterned interfaces. *Nat. Mater.* **2003**, *2*, 237–240. [CrossRef] [PubMed]
13. Ramos, S.; Charlaix, E.; Benyagoub, A. Contact angle hysteresis on nano- structured surfaces. *Surf. Sci.* **2003**, *540*, 355–362. [CrossRef]
14. Kim, D.; Hwang, W.; Park, H.C.; Lee, K.H. Super-hydrophobic nano-wire entanglement structures. *J. Micromech. Microeng.* **2006**, *16*, 2593. [CrossRef]
15. Cai, X.; Xiao, X. Research Progress of Superhydrophobic Surface Coatings. *Mod. Chem.* **2013**, *23*, 22–25.
16. Feng, L.; Song, Y.; Zhai, J.; Liu, B.; Xu, J.; Jiang, L.; Zhu, D. Creation of a super- hydrophobic surface from an amphiphilic polymer. *Angew. Chem. Int. Ed.* **2003**, *42*, 800–802. [CrossRef]
17. Ellinas, K.; Tserepi, A.; Gogolides, E. Superhydrophobic, passive microvalves with controllable opening threshold: Exploiting plasma nanotextured microfluidics for a programmable flow switchboard. *Microfluid. Nanofluidics* **2014**, *17*, 489–498. [CrossRef]
18. Tadanaga, K.; Morinaga, J.; Matsuda, A.; Minami, T. Super-hydrophobic–super- hydrophilic micropatterning on flowerlike alumina coating film by the sol–gel method. *Chem. Mater.* **2000**, *12*, 590–592. [CrossRef]
19. Youngblood, J.; McCarthy, T. Ultrahydrophobic polymer surfaces prepared by simultaneous ablation of polypropylene and sputtering of poly(tetrafluoroethylene) using radio frequency plasma. *Macromolecules* **1999**, *32*, 6800–6806. [CrossRef]
20. Liu, H.; Feng, L.; Zhai, J.; Jiang, L.; Zhu, D. Reversible wettability of a chemical vapor deposition prepared ZnO film between super-hydrophobicity and super- hydrophilicity. *Langmuir* **2004**, *20*, 5659–5661. [CrossRef]
21. Hirano, T.; Nakade, K.; Li, S.; Kawai, K.; Arima, K. Chemical etching of a semiconductor surface assisted by single sheets of reduced graphene oxide. *Carbon* **2018**, *127*, 681–687. [CrossRef]
22. Jiang, Y.; Wang, Z.; Yu, X.; Shi, F.; Xu, H.; Zhang, X.; Smet, M.; Dehaen, W. Self-assembled monolayers of dendron thiols for electrodeposition of gold nanostructures: Toward fabrication of super-hydrophobic/super-hydrophilic surfaces and pH-responsive surfaces. *Langmuir* **2005**, *21*, 1986–1990 [CrossRef] [PubMed]
23. Gao, N.; Geyer, F.; Pilat, D.W.; Wooh, S.; Vollmer, D.; Butt, H.J.; Berger, R. How drops start sliding over solid surfaces. *Nat. Phys.* **2018**, *14*, 191–196. [CrossRef]
24. Sarkiris, P.; Ellinas, K.; Gkiolas, D.; Mathioulakis, D.; Gogolides, E. Motion of Drops with Different Viscosities on Micro-Nanotextured Surfaces of Varying Topography and Wetting Properties. *Adv. Funct. Mater.* **2019**, *29*, 1902905. [CrossRef]
25. Lagubeau, G.; Le Merrer, M.; Clanet, C.; Quere, D. Leidenfrost on a ratchet. *Nat. Phys.* **2011**, *7*, 395–398. [CrossRef]
26. Ellinas, K.; Tserepi, A.; Gogolides, E. Durable superhydrophobic and superamphiphobic polymeric surfaces and their applications: A review. *Adv. Colloid Interface Sci.* **2017**, *250*, 132–157. [CrossRef]
27. Wu, C.; Ma, G.; Zhou, P. Research progress on boundary slip of fluid flow. *Adv. Mech.* **2008**, *38*, 265–282.
28. Hao, X.; Wang, L.; Ding, Y. Study on drag reduction of superhydrophobic surface. *Lubr. Eng.* **2009**, *34*, 25–28.
29. Poetes, R.; Holtzmann, K.; Franze, K.U. Steiner, Metastable underwater superhydrophobicity. *Phys. Rev. Lett* **2010**, *105*, 166104. [CrossRef]
30. Pan, S.; Kota, A.; Mabry, J.; Tuteja, A. Superomniphobic surfaces for effective chemical shielding. *J Am Chem Soc.* **2013**, *135*, 578–581. [CrossRef]
31. Ellinas, K.; Pujari, S.; Dragatogiannis, D.; Charitidis, C.; Tserepi, A. Zuilhof, Plasma micro-nanotextured, scratch, water and hexadecane resistant, superhydrophobic, and superamphiphobic polymeric surfaces with perfluorinated monolayers. *ACS Appl. Mater. Interfaces* **2014**, *6*, 6510–6524. [CrossRef] [PubMed]
32. Zhang, X.; Guo, Y.; Chen, H.; Zhu, W.; Zhang, P. A novel damage-tolerant superhydrophobic and superoleophilic material. *J. Mater. Chem. A* **2014**, *2*, 9002. [CrossRef]
33. Wang, H.; Liu, Z.; Wang, E.; Zhang, X.; Yuan, R.; Wu, S.; Zhu, Y. Facile preparation of superamphiphobic epoxy resin/modified poly(vinylidene fluoride)/fluorinated ethylene propylene composite coating with corrosion/wear-resistance. *Appl. Surf. Sci.* **2015**, *357*, 229–235. [CrossRef]

34. Zhu, X.; Zhang, Z.; Men, X.; Yang, J.; Wang, K.; Xu, X.; Zhou, X.; Xue, Q. Robust superhydrophobic surfaces with mechanical durability and easy repairability. *J. Mater. Chem.* **2011**, *21*, 15793–15797. [CrossRef]
35. Choi, C.; Kim, C. Large slip of aqueous liquid flow over a nanoengineered super-hydrophobic surface. *Phys. Rev. Lett.* **2006**, *96*, 066001. [CrossRef]
36. Li, Y.; Luo, B.; Guet, C.; Narasimalu, S.; Dong, L. Preparation and formula analysis of anti-biofouling titania–polyurea spray coating with nano/micro-structure. *Coatings* **2019**, *9*, 560. [CrossRef]
37. Huo, S.; Yu, Z.; Li, Y. Flow characteristics of water in superhydrophobic microchannels. *Chem. J. China* **2007**, *58*, 2721–2726.
38. Tsai, P.; Peters, A.M.; Pirat, C.; Wessling, M.; Lammertink, R.G.H.; Lohse, D. Quan-tifying effective slip length over micropatterned hydrophobic surfaces. *Phys. Fluids* **2009**, *21*, 112002–112008. [CrossRef]
39. Natrajan, V.; Christensen, K. The impact of surface roughness on flow through a rectangular microchannel from the laminar to turbulent regimes. *Microfluid. Nanofluidics* **2010**, *9*, 95–121. [CrossRef]
40. Hong, C.; Yamada, T.; Asako, Y.; Faghri, M. Experimental investigations of laminar, transitional and turbulent Gas flow in microchannels. *Int. J. Heat Mass Transf.* **2012**, *55*, 4397–4403. [CrossRef]
41. Arjun, K.; Rakesh, K. CFD analysis of thermal performance of microchannel nanofluid flow at different Reynolds numbers. *Songklanakarin J. Sci. Technol.* **2019**, *41*, 109–116.
42. Gatapova, E.; Dmitry, S.G. The Drag Reduction of Microchannel Flow by Contrast Wettability. *MATEC Web Conf.* **2016**, *72*, 4. [CrossRef]
43. Huang, J.; Deng, Y.; Yi, F.; Peng, L. Experimental and numerical investigation on thin sheet metal roll forming process of micro channels with high aspect ratio. *Int. J. Adv. Manuf. Technol.* **2019**, *100*, 117–129. [CrossRef]
44. Madavan, N.; Deutsch, S.; Merkle, C. Measurments of local skin friction in a microbubblemodified turbulent boundary layer. *J. Fluid Mech.* **1985**, *156*, 237–256. [CrossRef]
45. Byun, D.; Kim, J.; Han, S.; Hoon, C. Direct measurement of slip flows in superhydrophobic microchannels with transverse grooves. *Phys. Fluids* **2008**, *20*, 113601. [CrossRef]
46. Deutsch, S.; Castano, J. Microbubble skin friction reduction on an axisymmetric body. *Phys. Fluids* **1986**, *29*, 3590–3596. [CrossRef]
47. Zhou, Y.; Chang, H. Numerical simulation of hydrodynamic and heat transfer characteristics of slug flow in serpentine microchannel with various curvature ratio. *Int. J. Heat Mass Transf.* **2019**, *55*, 3343–3358. [CrossRef]
48. Papageorgiou, D.P.; Tsougeni, K.; Tserepi, A.; Gogolides, E. Superhydrophobic, hierarchical, plasma-nanotextured polymeric microchannels sustaining high-pressure flows. *Microfluid. Nanofluid.* **2013**, *14*, 247–255. [CrossRef]

coatings

Article

Non-Fluorinated, Sustainable, and Durable Superhydrophobic Microarrayed Surface for Water-Harvesting

Oriol Rius-Ayra *, Sheila Fiestas-Paradela and Nuria Llorca-Isern

CPCM Departament de Ciència dels Materials i Química Física, Facultat de Química, Universitat de Barcelona, Martí i Franquès 1, 08028 Barcelona, Spain; shfiestp7@alumnes.ub.edu (S.F.-P.); nullorca@ub.edu (N.L.-I.)

* Correspondence: oriolriusayra@ub.edu

Received: 2 March 2020; Accepted: 24 March 2020; Published: 26 March 2020

Abstract: Water scarcity is a worldwide issue that significantly affects the environment, population, and economy of the arid zones. In this study, we report a straightforward method for water-harvesting based on modifications of the surface wettability. Using magnesium chloride, lauric acid, and electrodeposition process, a superhydrophobic surface (155°) is obtained. Morphological characterization techniques allow determination of the characteristic flower-like microstructures combined with close packed nanoarrays that lead to the hierarchical structure. Furthermore, the coating presents vertically aligned microarrays in a non-linear cone morphology formed by dynamic templating of hydrogen bubbles. From a chemical point of view, magnesium laurate is responsible for the surface tension decrease. To determine the durability of the obtained surface ultra-violet (UV) light test and abrasive paper test, tests are carried out revealing high durability against these severe conditions. The water-harvesting ability of the superhydrophobic surface is studied at 45° and 90° tilted samples. The capacity of the water to be harvested efficiently is found to be at 90° tilt under fog conditions. The use of green reactants associated with this hierarchical structure broadens a new scope for sustainable freshwater collection and it becomes an excellent example of a green solution.

Keywords: non-fluorinated; superhydrophobic; water-harvesting; fatty acid; robust; durable

1. Introduction

Water shortages and scarcity is a worldwide phenomenon that significantly alters the environment as well as the population in arid zones. Moreover, access to safe drinking water remains a challenge in several countries, which leads to the development of new strategies to achieve a new scenario in which materials used must be remarkably sustainable.

Indeed, surfaces with extraordinary wettable properties are attracting attention due to the capability of collecting water from the environment. In fact, surfaces with a water contact angle (WCA) higher than 150° as well as a contact angle hysteresis (CAH) ≤ 10° are defined as superhydrophobic [1,2]. Because of their extraordinary surface properties, novel applications are emerging to solve environmental problems such as oil/water separation in oil spills [3–7], corrosion resistance in marine conditions [8,9], or reducing ice adhesion in cold environments [10,11]. Moreover, water-harvesting in some environments can contribute to solving global issues. Because of that, several complex methods are carried out to collect water droplets by combining superhydrophobic–hydrophilic surfaces such as polymerization [12], templating [13], biotemplating [14] or nanoimprint lithography [15]. Despite the results showing these references, the reactants used to decrease surface tension such as trimethylchlorosilane (TMCS) or perfluorooctyltriethoxysilane (FAS) are combined with solvents such as chlorobenzene or chloroform that are extremely harmful for the environment and human health. Therefore, the development of innovative and sustainable strategies [16–18] appears to be necessary to generate a new framework for superhydrophobic surfaces. Environmentally friendly materials [19–22] are a forward-looking approach towards materials science because they also play an important role in applications to distinct ecosystems due to their benign properties. For that reason, eco-friendly surfaces without the presence of fluorine compounds remain a challenge [23,24]. Indeed, long-chain fatty acids [25,26] play a key role in this subject because of their intrinsically hydrophobic character. In this direction, different strategies can be found in the literature such as liquid phase deposition (LPD) of cinnamic acid from cinnamon or myristic acid from nutmeg to prepare superhydrophobic copper surfaces [27]. Others rely on the co-precipitation method used to obtain Fe_3O_4 magnetic nanoparticles (MNPs) via an eco-friendly route [28]. Moreover, superhydrophobic polydimethylsiloxane (PDMS) monolith has also been prepared by emulsion templating process [29], or biomass-derived banana peel aerogel from the combination of different techniques allowing separation of oil-in-water systems [30].

In the present research, a new sustainable surface is obtained by electrodeposition of lauric acid with magnesium chloride that leads to magnesium laurate as a main chemical compound responsible for surface tension decrease. This effect combined with the flower-like structures conduced superhydrophobic properties observed (WCA = 155°). To sum up, the as-prepared surface is highly durable and is still superhydrophobic after several cycles against UV light exposure or abrasive test. Finally, the completely superhydrophobic surface is composed of several microarrays in a non-linear cone shape caused by dynamic templating during electrodeposition, which allows water condensation under fog conditions and collects water droplets for human consumption.

2. Materials and Methods

2.1. Electrodeposition

The present methodology is based on electrochemical deposition (EDP) that has been carried out on aluminum substrate UNS A91070 thin plate (10 × 40 × 1 mm) (Servei Estació, Barcelona, Spain) previously grinded with P1200 SiC abrasive paper to clean and remove impurities from the substrate surface. The electrolytic bath is prepared as follows: the reactants, 0.1 M lauric acid (Scharlab, S.L., Barcelona, Spain) and 0.05 M $MgCl_2$ (Panreac, Barcelona, Spain), are dissolved in a solution of absolute ethanol (Scharlab, S.L., Barcelona, Spain). Two aluminum substrates (acting as electrodes) are immersed vertically in the previous solution, which works as the electrolytic bath, and both plates are separated 20 mm from each other. After applying a current density of 0.02 A/dm^2 for 900 s, the superhydrophobic system is deposited onto the cathode. After the electrodeposition, the samples are removed, cleaned with ethanol, and dried in a fume hood for a few seconds.

2.2. Durability Tests

Two different tests are carried out to determine the durability of the as-prepared superhydrophobic surface. On the one hand, the surface is exposed under UV light (λ = 250 nm) using a Philips TUV 8W G8T5 UV lamp with a 25 cm distance between the lamp and the sample. The test allows evaluation of the stability of the surface after 300 min of exposure. On the other hand, a sandpaper abrasive test is also used to determine the surface durability in severe abrasive conditions. The superhydrophobic surface is placed in contact with SiC P1200 sandpaper and it is moved 10 cm across for 10 cycles while a pressure of 5 kN/m^2 is applied. After each abrasion cycle, the WCA as well as the sliding angle (SA) are measured.

2.3. Water-Harvesting Test

To study the water-harvesting ability of the as-prepared sustainable surface, a conventional controlled salt fog chamber (ISO 9227) was set up at room temperature (25 ± 1.1 °C) and a working pressure of deionized water of 172 kPa. All the samples had an area of 10 × 4 mm and were exposed for 2 h under fog conditions; the weight of the collected water was measured. In addition, the samples were held at two different tilt angles (45° and 90°) to determine any effect in water-harvesting rate.

2.4. Characterization Techniques

Different characterization techniques were used to determine morphological structure as well as the chemical composition of the obtained surface. The sample surface was characterized using FE-SEM JEOL J-7100 electron microscope (JEOL Ltd., Tokio, Japan) in order to study its detailed morphology and energy dispersive X-ray spectroscopy detector (EDS) (Oxford Instruments, Oxfordshire, England) to determine the elemental composition of the surface. Atomic Force Microscopy (AFM) was used to analyze the topographic structure using AFM Multimode 8 Microscope with electronic Nanoscope V (Bruker, Billerica, MA, USA) in Peak Force tapping Mode with a tip Sharp Nitride Lever (SNL) from (Bruker, Billerica, MA, USA) and the cantilever with resonance frequency of 65 kHz and a spring constant of 0.35 N/m. The study of the surface nanostructure was performed with a transmission electron microscope, JEOL JEM-2100 (TEM) (JEOL Ltd. Tokio, Japan) with EDS detector for microanalysis; the samples were supported onto a holey carbon film on a Cu grid. The roughness of the surface was measured by confocal microscopy using a LeicaScan DCM3D on a surface of 1270 × 950 μm^2 using a white light beam. X-Ray Photoelectron Spectroscopy (XPS) (PHI, Chanhassen, MN, USA) was carried out to determine chemical composition by PHI ESCA-5500 using a monochromatic X-Ray source (Kα (Al) = 1486.6 eV and 350 W); Multipak software was used to perform deconvolutions and calculate the atomic percentage of each element. Attenuated Total Reflectance Fourier Transformed Infrared (ATR-FTIR) spectroscopy was also used to establish the presence of hydrocarbon acid and its chemical bonds; for this purpose, a Fourier Bomem ABB FTLA in the range 4000–525 cm^{-1} at a resolution of 4 cm^{-1} was used. μ-Raman spectroscopy analysis was carried out to determine chemical composition of the obtained surface using a Jobin–Yvon LabRam HR 800 dispersive spectrometer, coupled to an optical microscope Olympus BXFM (Horiba, Kioto, Japan). The CCD detector was cooled at –70°C. Laser lines were 532 nm in a spectral range of 4000–100 cm^{-1}, dispersive gratings at 600 l/mm and at 5 mW power. Phase characterization of the surface was obtained by X-Ray Diffraction (XRD) from PANalytical X'Pert PRO MPD Alpha1 powder diffractometer in Bragg-Brentano $\theta/2\theta$ geometry of radius 240 millimeters and Cu Kα1 radiation (λ = 1.5406 Å). Contact angle was measured using a digital microscope Levenhuk and 3.5 μL of deionized water at room temperature. The reported WCA values are the average of three measurements of droplets at different parts on the surface. SA were measured when the samples were tilted until water droplets rolled off and CAH were measured by increasing and decreasing droplet volume.

3. Results and Discussion

3.1. Structural Characterization

Field-Emission Scanning Electron Microscopy (FE-SEM) observations were performed to study the surface morphology. After 900 s of EDP, the aluminum substrate surface was completely covered with micro- and nano-structures. The combination of both features causes a hierarchical structure adopting a flower-like morphology that plays a key role in the modification of the surface wettability. These characteristic structures are built up in two distinct levels: the microstructure shown in Figure 1a, formed itself by the nanolayers as seen in Figure 1b.

EDS microanalyses (Figure 1c) of the nanostructure indicated the presence of a carbon peak corresponding to lauric acid and a contribution from the previous carbon sputtering (Kα = 0.277 keV); also, there was a less intensive peak of magnesium appeared in the spectra belonging to the obtained surface layer (Kα = 1.253 keV). Oxygen (Kα = 0.525 keV) and aluminum (Kα = 1.486 eV) are also present in the EDS spectra, both from the substrate as well as the inherent alumina to the aluminum substrate surface. A certain contribution of the substrate and lauric acid ($C_{12}H_{24}O_2$) are confirmed.

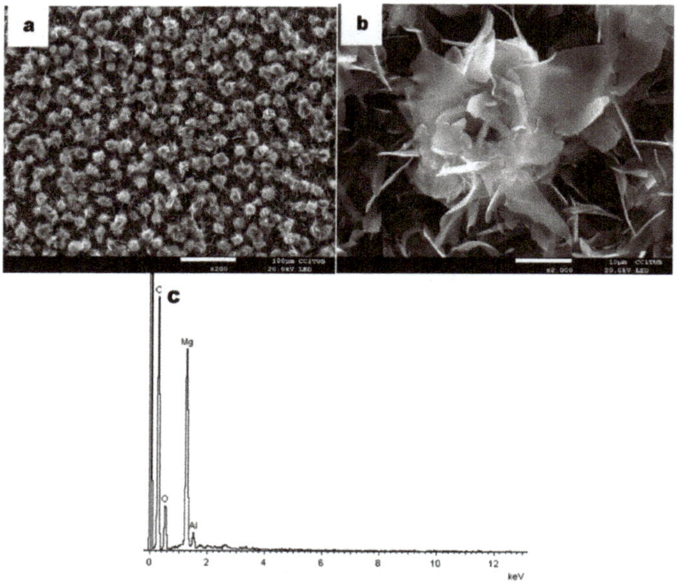

Figure 1. FE-SEM images of the as-prepared superhydrophobic (SH) surface on aluminum substrate: (**a**) distribution of microstructures on the substrate after 900 s of electrodeposition; (**b**) at higher magnification the flower-like structure and its nanostructured layers are observable; and (**c**) EDS microanalysis where Mg and C from the obtained coating and Al from substrate can be identified.

AFM is used as a high-performance imaging technique allowing more accurate analysis of the hierarchical structure of the surface at the nanometer scale. From the AFM observation, the surface of nanolayers is found to be highly organized by lamellar nano-structures in which the smooth layers overlap (Figure 2a,b bottom left) and are highly stacked to each other as can be seen in Figure 2c,d. The layer thickness was measured and was 32.8 ± 2.9 nm.

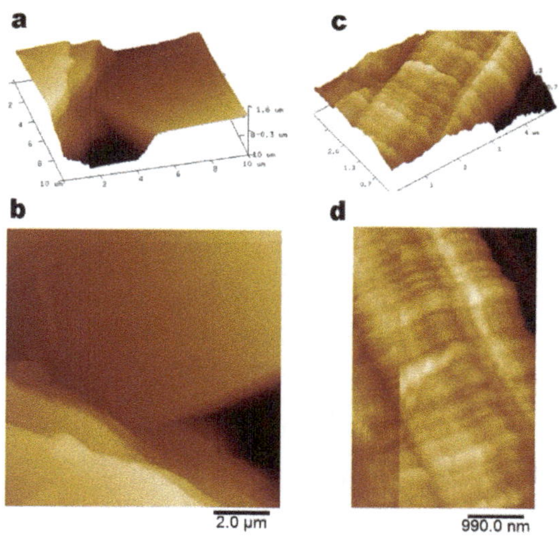

Figure 2. AFM images showing lamellar nano-structures stacked on each other in two different areas: 10 × 10 µm (**a**) (**b**) and 4.0 × 3.4 µm (**c**) (**d**).

High-resolution transmission electron microscopy (HR-TEM) allows observation of the structure at the nanometer scale as well as determination of the elemental composition by EDS microanalysis of the as-prepared SH coating. The structure observed corresponds to the unit (petal) of the flower-like morphology made of thin layers that overlap themselves (Figure 3a) and planes in the layer itself (Figure 3b). These units of the coating do not contain long-range order in the atomic lattice, giving diffuse ring diffraction patterns in the corresponding selected-area electron diffraction (SAED) obtained. The weak diffraction rings in Figure 3c suggest certain arrangements that can be considered crystalline in nature of the petal morphology, consistent with a distinguishable diffraction ring that corresponds to a monoclinic crystal structure according to the Joint Committee on the Powder Diffraction Standards (JCPDS) with data card 8-528. Figure 3d shows the EDS microanalysis in which the magnesium associated with the SH coating is identified (Kα = 1.253 keV). A peak of aluminum from the substrate (Kα = 1.486 keV) and an intense peak of carbon (Kα = 0.277 keV) from the lauric acid and also from the carbon film of the copper grid (Lα = 0.930 keV and Kα = 0.930 keV), chloride peaks (Kα = 2.621 keV) are assigned to the presence of magnesium chloride from the reactant used.

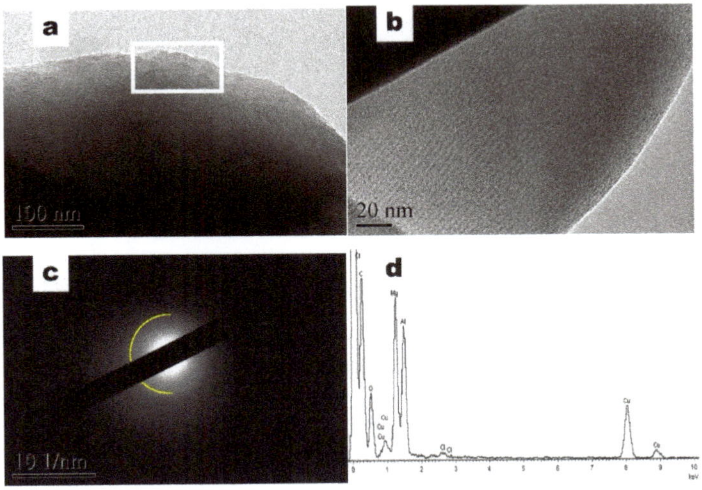

Figure 3. HR-TEM images of the as-prepared coating that is structured in nanolayers that overlap themselves (**a**), (**b**); (**c**) the SAED pattern showing weak rings from nanolayers; and (**d**) the EDS spectrum shows that the composition is mainly Mg and C (as the TEM grid is Cu and Cl are associated with the $MgCl_2$ reactant used).

XRD technique should allow determination of the crystalline character of the surface. As shown in Figure 4, the coating presents five peaks assigned to the aluminum substrate and corresponds to the planes 2θ = 38.5° (111), 2θ = 44.7° (200), 2θ = 65.1° (220), 2θ = 78.2° (311), and finally 2θ = 82. 4° (222). Additionally, a reflection at 2θ = 20.798° is distinguishable (Figure 4 inset) that is characteristic of the lamellar structures of organic compounds arranged in a monoclinic set system that can be assigned to the space group $P2_1/a$ [31–34]. Hence the superhydrophobic surface is made of well-organized layers.

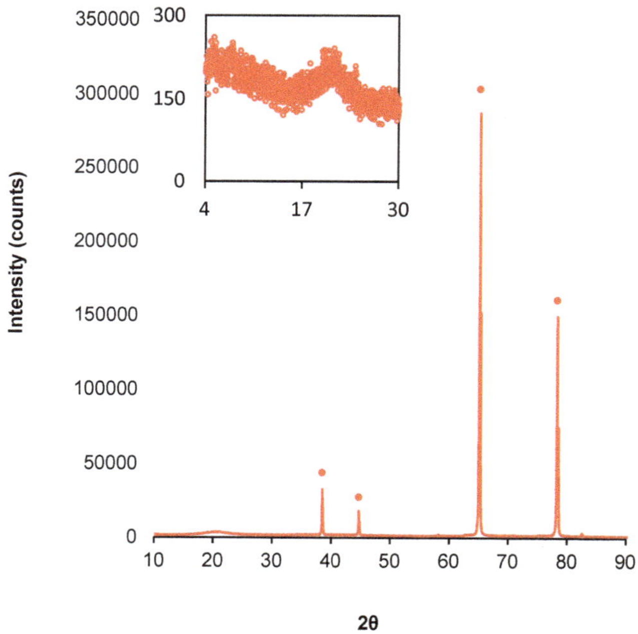

Figure 4. XRD peaks in the inset indicate well-organized layers from an organic lamellar structure at $2\theta = 20.798°$. Peaks from the aluminum substrate (•) are indicated in the diffractogram.

For structural characterization, let us interpret the surface morphology as the combination of each flower-like feature with nanolayers giving rise to the hierarchical structure. The assembly of these hierarchical structures increases the roughness allowing the air to be trapped in the whole surface pockets producing a Cassie–Baxter wetting regime [35]. In addition, the nanolayers of the superhydrophobic surface are arranged in a monoclinic crystal structure that can be related to greater lateral Van der Waals interactions between laurate chains in the lattice [32].

3.2. Chemical Characterization

Fourier transform infrared spectrometry (ATR-FTIR) is used to determine the chemical composition related to lauric acid as well as to the new eco-friendly coating (Figure 5). Pure lauric acid [36–38] and the formed coating presents different common bands, which are sharp peaks at ca. 2951 cm^{-1}, 2911 cm^{-1} and 2845 cm^{-1} that correspond to stretching of the alkyl chain of $\nu_{as}CH_3$, $\nu_{as}CH_2$, ν_sCH_2-CH_2. At ca. 1470 cm^{-1} weak and broad bands assigned to δCH_2 and δH–C–H bendings can also be found. Finally, both samples coincide in three weak and sharp bands situated at ca. 1122 cm^{-1} that correspond to δCH bending out of plane, ca. 1088 cm^{-1} CH_3 symmetric deformation and ca. 776 cm^{-1} from the vibration of δC-H bending out of plane. For SH coating, the vibration with sharp and very strong band at ca. 1696 cm^{-1} assigned to νC = O stretching is no longer present in the coating [39,40]. Moreover, sharp and medium intense band at ca. 1300 cm^{-1} is assigned to νC–O stretching and two sharp and medium intense bands at ca. 1192 cm^{-1} and ca. 934 cm^{-1} are assigned to the stretching vibration of –OH group. In contrast, superhydrophobic coating presents three distinct sharp and weak bands at ca. 1559 cm^{-1}, ca. 1442 cm^{-1} and ca. 1408 cm^{-1} that are assigned to the vibrations of $\nu_{as}COO^-$, $\nu_{as}COO^-$, and ν_sCOO^- stretching, respectively. This characteristic difference indicates that no free lauric acid on the surface is left after the electrodeposition and a new functional group is formed assigned to carboxylate formation (COO^-).

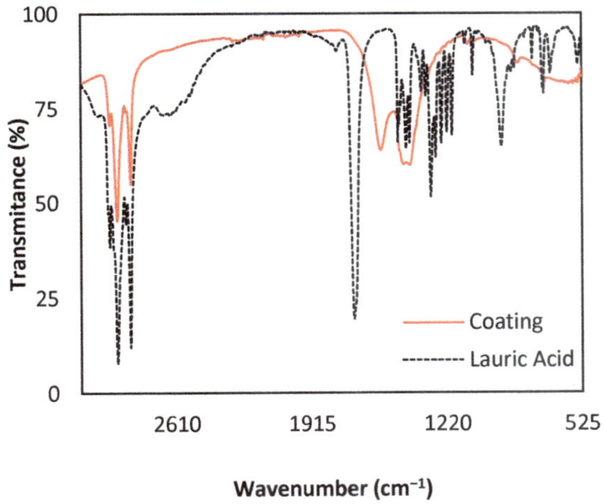

Figure 5. ATR-FTIR spectroscopy technique allows determining the formation of carboxylate compound after the electrodeposition process.

μ-Raman spectroscopy was used to complement ATR-FTIR and determine the chemical composition and consequently the presence of laurate. This characterization technique is particularly important because it reveals the characteristic fingerprint for organic compounds due to alkyl chain [41,42]. Figure 6 depicts the Raman spectrum of the obtained superhydrophobic surface where four types of group bands can be identified. The first region below 1000 cm^{-1} corresponds to low-frequency bands and are assigned to γCH_2 rocking vibration [43,44]. The group of bands between 1064 cm^{-1} and 1123 cm^{-1} are assigned to symmetric ν_sC–C symmetric stretch [45]. Bands between 1295 cm^{-1} and 1446 cm^{-1} correspond to δCH_2 and δCH_3 bending [44,46]. At 2735 cm^{-1} can be found a band also assigned to δCH_2 bending. Finally, ca. 2900 cm^{-1} this group of strong bands are assigned to νC–H stretch [43]. Those bands are characteristic of lauric acid derivative such as magnesium laurate.

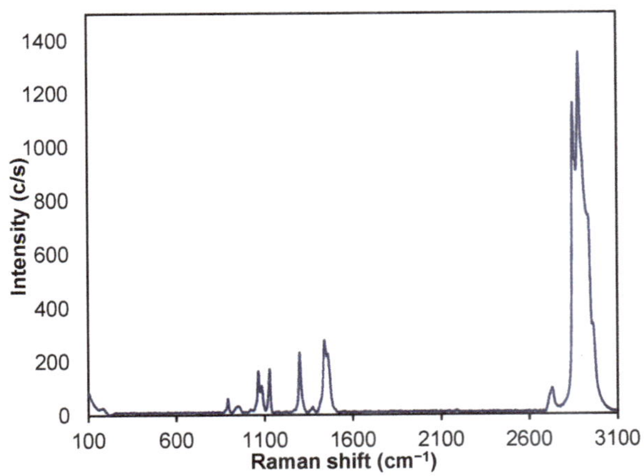

Figure 6. Raman spectrum of superhydrophobic coating assigned to laurate because of the fingerprint of alkyl chain.

Comparing the spectra of Raman with the ATR-FTIR, it can be found that weak bands assigned in ATR-FTIR such as C–H vibrations are strong in Raman spectrum and vice versa, which is a common behavior between both spectroscopic techniques. This particularity allows determination of the chemical composition of the as-prepared superhydrophobic surface assigned to laurate group.

High-Resolution (HR) XPS was used to investigate the chemical composition as well as the chemical environment of the obtained superhydrophobic coating. The HR-XPS spectra shows C-1s, O-1s, and Mg-2s of the coating after 900 s of EDP and the atomic concentration is 77.65% of C-1s, 18.06% for O-1s, and 1.998% of Mg-2s. It is worth noting the presence of Ca-2s (1.01 %), which is considered an impurity of the reactants. The peak in the C-1s region can be deconvoluted in three distinct components (Figure 7 a) corresponding at 282.7 eV to C-C bond, 285.6 eV to carbonyl group (C = O) form the carboxylate and finally 289.2 to carboxylate group (O–C = O) [47,48]. The O-1s spectra is deconvoluted to three peaks (Figure 7b) assigned to C = O at 530.1 eV, C-O bond at 532.7 eV and O–C = O from the carboxylate group at 535.2 eV bonded to a metallic atom [48,49]. For Mg-2s spectrum the deconvolution (Figure 7 c) presents a peak at 90.57 eV from Mg (II) oxidation state [50].

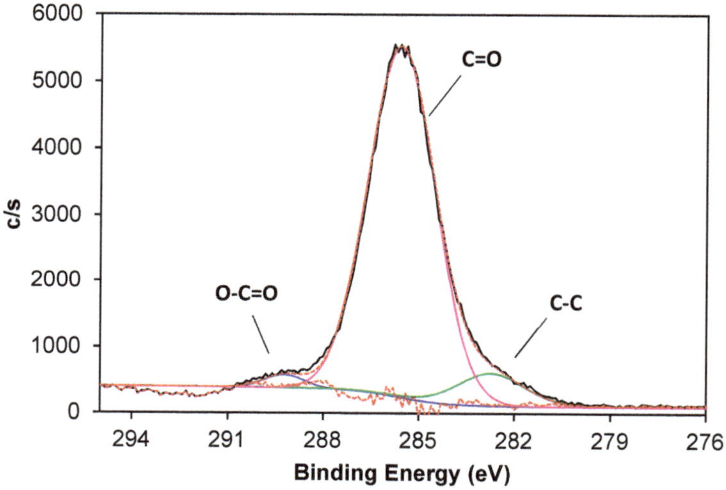

(**a**)

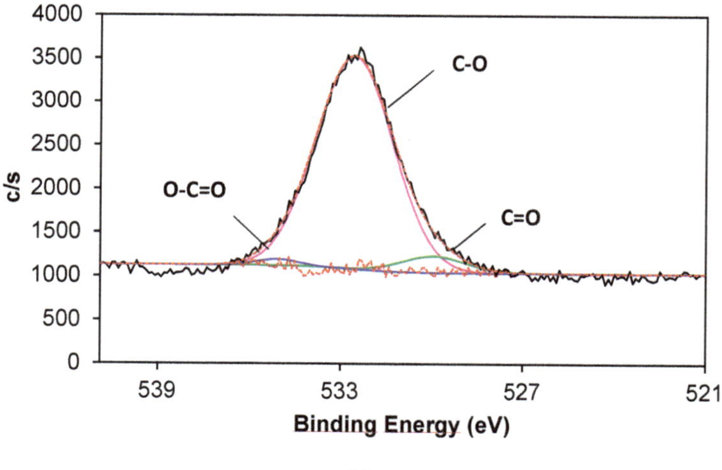

(**b**)

Figure 7. *Cont.*

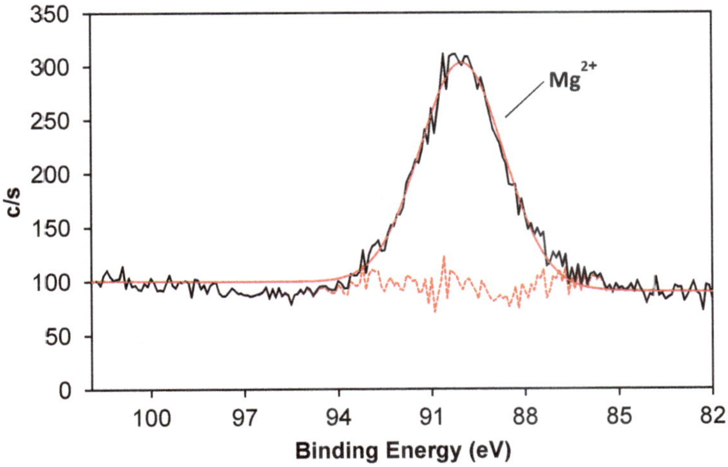

(c)

Figure 7. HR-XPS confirms the generation of a magnesium carboxylate after the EDP that decreases the surface tension of the as-prepared coating (**a**) C-1s, (**b**) O-1s and (**c**) Mg-2s.

The combination of these spectroscopic techniques allows the determination of the chemical composition of the as-prepared superhydrophobic surface. Taking into account the ATR-FTIR, it shows the characteristic bands of carboxylate group after EDP of lauric acid combined with $MgCl_2$ while in Raman spectrum the fingerprint of laurate alkyl chain is found. Moreover, O-1s and C-1s HR-XPS spectra reveals the presence of carboxylate functional group bonded to a metallic atom that is assigned to magnesium (II) as depicted in Mg-2s spectrum. Taking into account these results, the main component of the system is assigned to magnesium laurate ($Mg(C_{11}H_{20}COO)_2$) that contributes to decrease the surface tension of the final system.

From an electrochemical point of view, near the aluminum surface, magnesium (II) ions react with lauric acid while it releases H^+ to the hydro-alcoholic medium that tends to increase its concentration locally [51,52].

$$Mg^{2+} + 2C_{11}H_{20}COO^- \rightarrow Mg(C_{11}H_{20}COO)_2$$

Water is oxidized onto the anode while magnesium laurate is formed on to the cathode where at the same time it also takes place hydrogen evolution reaction (HER):

$$H_2O_{\ (l)} \rightarrow O_{2\ (g)} + 4H^+_{(aq)} + 4e^-$$
$$2e^- + 2H^+_{(aq)} \rightarrow H_{2\ (g)}$$

Therefore, meanwhile, magnesium laurate flower-like microstructures are formed through the whole surface of aluminum substrate, and hydrogen bubbles promote the obtained coating, itself vertically aligned in microarrays due to its detachment and vertical flow, which is an indication that hydrogen bubbles perform as a dynamic template. The obtained morphology of microarrays is characterized using confocal microscopy, as shown in Figure 8.

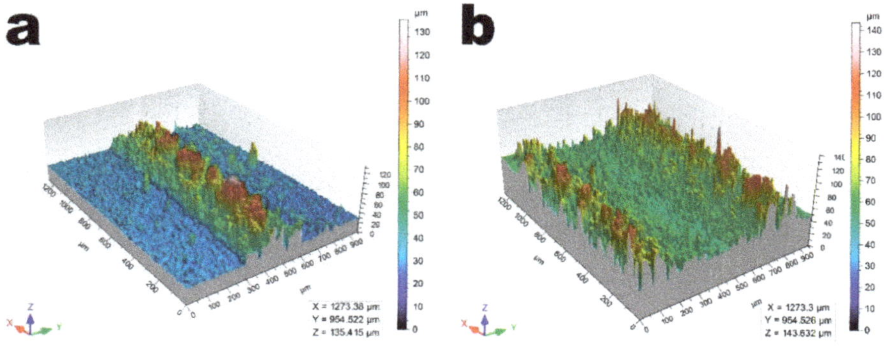

Figure 8. Confocal microscopy images show: (**a**) the formed microarrays in a non-linear cone shape and (**b**) pitch between cones. These features were caused by dynamic templating of hydrogen bubbles during electrodeposition of magnesium laurate.

The morphology of microarrays can be defined as a non-linear cone shape with 20° tip angle, 285.4 ± 22.5 μm base length, and height 21.7 ± 7.3 μm while lower tip angle is 57° (Figure 8a). Additionally, the distance between two consecutive non-linear cones, known as pitch, is 795.9 ± 5.6 μm (Figure 8b).

3.3. Wettability Properties

Superhydrophobicity is precisely defined by several parameters related to the angle between a surface and a water droplet such as WCA, SA, and CAH [2,53]. In our case, after 900 s electrodeposition process the WCA = 155° ± 0.6° and the as-prepared surface shows superhydrophobic properties and the SA is also measured and results 5° ± 0.1° (Figure 9a,b). Furthermore, CAH = 1 ± 0.5°. Considering these results, it can be concluded that water droplet roll-off across the surface.

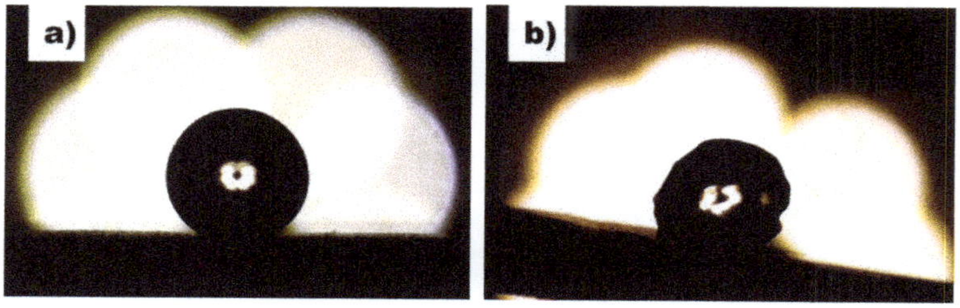

Figure 9. Images of contact angle measurements: (**a**) WCA = 155 ± 0.6° and (**b**) SA = 5 ± 0.1°. Both characteristics leads to a superhydrophobic surface with self-cleaning properties.

Moreover, it is important to study the behavior of these coatings under different conditions to understand and determine their durability. Hence, UV light resistance test [54] is carried out to determine the capability of the superhydrophobic surface to resist environmental conditions. After 300 min of UV light exposure (Figure 10), there was no remarkable difference in the WCA between the coating before the test and after 300 min under UV light exposure. Indeed, the WCA remained higher than 150° after every exposure leading to a durable superhydrophobic surface in these conditions. Regarding the SA, it slightly increased from 5° to 7° and because the value was still lower than 10° after the UV tests, it indicates the robustness of the SH surface under UV light conditions.

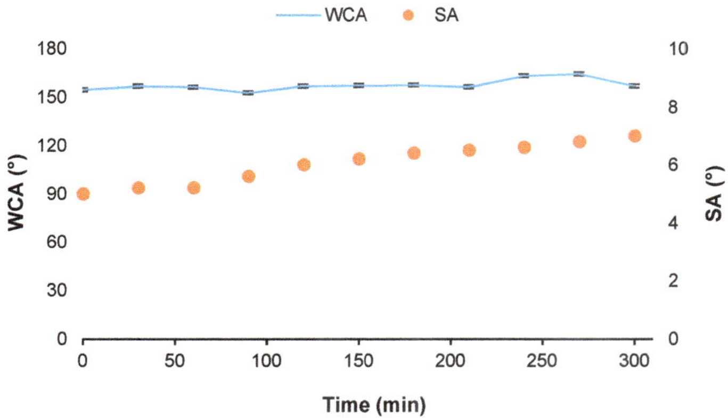

Figure 10. After several cycles of UV light test the WCA is not affected at all, which denotes a high stability of the coating under energetic light. SA increases slightly but it still presents self-cleaning properties after 300 min under UV light exposure.

Taking into consideration the resistance against severe mechanical test, a sandpaper abrasion test (SiC P1200) was performed because it is an effective method to evaluate the durability against mechanical abrasion of superhydrophobic surfaces. The test consists of forcing the contact between the surface and the abrasive paper with an applied pressure of 5 kN/m^2 through 10 cm, this test is repeated 10 times. After the first abrasive cycle, a white powder could be found on the sandpaper although the WCA remains higher than 150° in the coated sample. Despite this residue and after 10 cycles (Figure 11a), the surface is still superhydrophobic and the corresponding WCA was also measured with a value of 152° and SA = 6°. The results indicate high durability of the SH surface and maintains its self-cleaning properties despite the sigmoidal increase after the fifth cycle (Figure 11a) caused by a decrease in roughness. Furthermore, self-cleaning properties of superhydrophobic surface after abrasion test were also investigated. Typically, a layer of soot was poured over the surface, which was tilted 6°, and the water droplet easily slides along the surface while soot powder was pick up and the surface was completely cleaned (Figure 11b). This result reveals that despite mechanical abrasion against the surface, it retains high WCA and low SA where both features are responsible for low adhesion and self-cleaning properties.

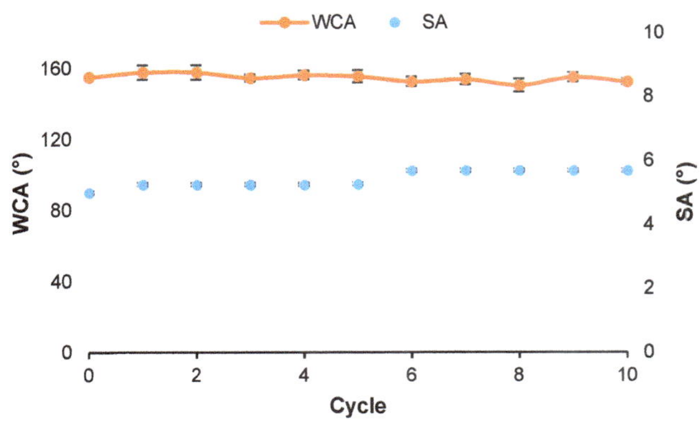

(a)

Figure 11. *Cont.*

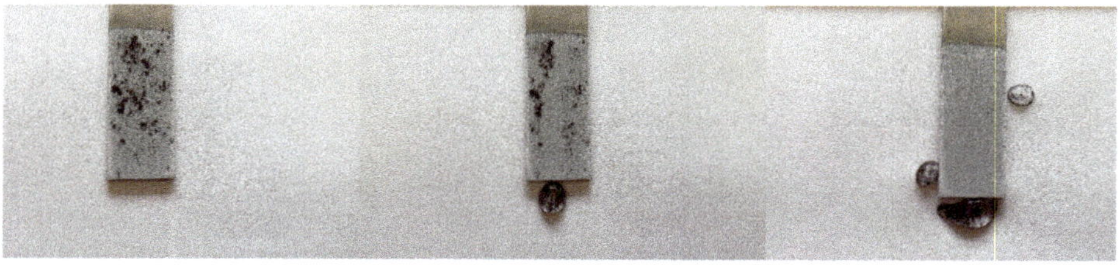

(**b**)

Figure 11. Durability of the as-prepared coating is showed by (**a**) WCA remains slightly constant and higher to 150° after each sandpaper abrasive cycle while SA increases until a maximum value of 6°; and (**b**) self-cleaning properties were investigated after abrasive paper test.

As the roughness of the surface samples plays a key role to define the superhydrophobic character [1], RMS (root-mean-square) is measured using confocal microscopy. Before EDP, aluminum substrate shows RMS = 0.52 ± 0.16 µm (Figure 12a) whereas after that process, the superhydrophobic surface takes a value of 8.59 ± 0.39 µm (Figure 12b). An increase in the roughness of the surface will lead to a rise in the superhydrophobic character of the as-prepared surface. The results of RMS values of the as-prepared coating are compared with the measured RMS after sandpaper abrasion test where RMS takes a value of 2.99 ± 0.76 µm (Figure 12c).

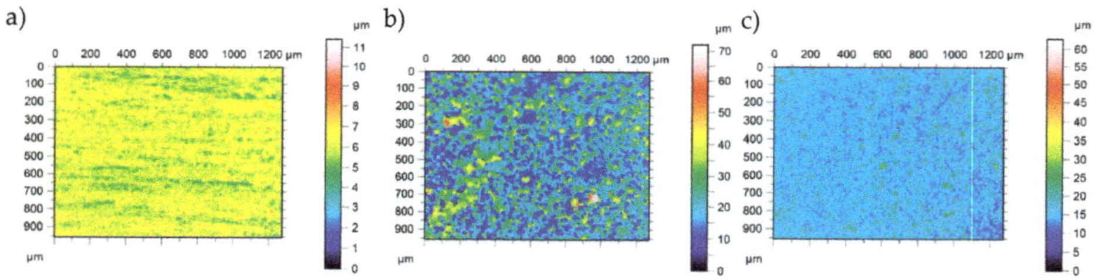

Figure 12. RMS values measured by optical confocal microscopy are aluminum substrate (**a**) 0.52 ± 0.16 µm; superhydrophobic surface (**b**) 8.59 ± 0.39 µm and after 10 cycles of abrasion test (**c**) 2.99 ± 0.76 µm. The EDP process increases the roughness while the abrasion test reduces it.

The difference between the RMS values before and after the sandpaper abrasive test is due to its ability to smooth the surface. Despite decreasing the value of roughness, the superhydrophobic character is not severely affected (WCA = 152°) and it remains superhydrophobic. To better comprehend the resistance against abrasive paper test and evaluate the durability of the as-prepared surface, FE-SEM is used to determine the morphology of the surface after 10 cycles and EDS allows identification of the semiquantitative composition before and after the test. It can be seen that the flower-like structure (Figure 13a) is no longer present at the upper structures and in fact they are smoother compared with the previous morphology (Figure 1). Despite this, several bumpy structures are present and composed of a hierarchical structure. In addition, the elemental composition of the surface after the abrasive paper test is determined and compared using EDS microanalyses (Figure 13b). The spectrum shows that the composition is the same compared to the coating before abrasive test (Figure 1c) but the peak assigned to Mg is less intense than the coating itself before abrasion. These results demonstrate that the composition does not change significantly after the coating surface is damaged by abrasive paper test. This illustrates that the hierarchical structure of the as-prepared coating is chemically homogeneous.

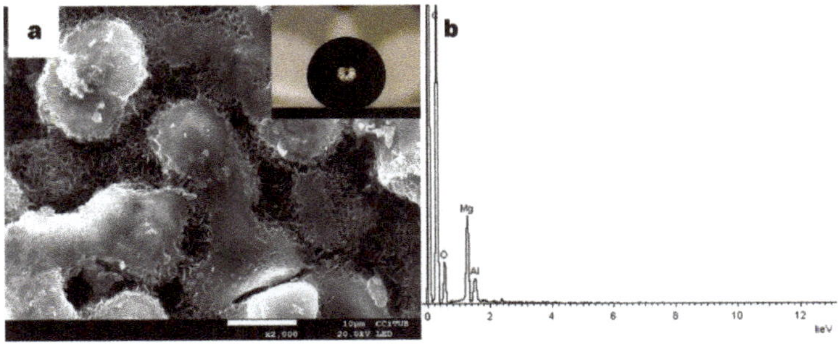

Figure 13. FE-SEM micrograph and EDS spectra shows (**a**) the coating after 10 cycles of abrasive paper test with a smooth structure and (**b**) similar elemental composition after the test is compared to Figure 1c.

One of the most problematic consequences of wear in superhydrophobic surfaces is the loss of character hence moving to hydrophilic surface due to the generation of pinning sites through the whole surface. In fact, due to the homogeneity of flower-like structures of these samples, which are made of the hydrophobic compound magnesium laurate, wear will not introduce hydrophilic character to the as-prepared coating [55–57]. Moreover, as the hierarchical structure is not completely damaged and still present on the surface, and the WCA and the superhydrophobic behavior of the surface is not severely affected. Indeed, the combination of both characteristics, chemical composition homogeneity and hierarchical structure, keeps the Cassie–Baxter wetting state of the coating that promotes superhydrophobicity to the surface. Furthermore, roughness loss does not severely affect its wettability properties, which leads to a durable superhydrophobic surface that also presents self-cleaning properties.

3.4. Water-Harvesting

To understand the capability of the superhydrophobic coating to collect water droplets from the fog/mist environment, the experimental set-up is carried out in a conventional fog chamber. Water collection rate (ξ) was calculated following Equation (1) at different times ranging from 10 min to 120 min:

$$\xi = m_{H2O}/S{\cdot}\mathrm{t} \quad (1)$$

where m is the water mass measured when it is collected, S is the total surface of the coating on to aluminum substrate, and t is time. Samples are placed vertically (90°) and with a tilt (45°) to study the influence of the position in water-harvesting. Figure 14 depicts collecting water rate of coated samples (45° and 90°) and two aluminum substrates as references in both positions (tilt 45° and vertical 90°). As shown, both aluminum references present a water collection rate near zero after 120 min of exposure (6.3×10^{-5} µg·mm^{-2}·min^{-1} in case of 90°, and 3.7×10^{-5} µg·mm^{-2}·min^{-1} for sample tilted 45°), it indicates that these samples are not capable of collection water due to the hydrophilic character of aluminum substrate (WCA = 83 ± 0.9°). Thus, water droplets are strongly adhered on the substrate and cannot slide through the surface to be collected. In the case of coated samples, collection rate increases in both positions but the sample in the vertical position is more efficient; indeed, after 2 h of water collection, the 45° tilt sample allows collection of 9 µg·mm^{-2}·min^{-1} while the vertical sample collects 13 µg·mm^{-2}·min^{-1}. Furthermore, in the case of the tilt sample, two different stages of water collection can be identified: the first one from the beginning until it reaches a maximum at 30 min and then remains flat until 60 min of fog test; here, it can be seen that 45° collection rate is higher than the rate of 90° sample (before 40 min); this difference is because at 45° water droplets are slightly retained, which tends to increase their volume and increase the amount of collected water. After this, the second stage behaves in a similar way: water collection rate increases until 90 min and then remains constant until

120 min of fog exposure. This specific behavior is caused by the tilt angle: initially water is collected at a proportional rate for the first 30 min. After that, collection rate remains constant, which indicates a decrease of water collected mass until 60 min of fog exposure. Then, this behavior is repeated with an increase of collected water. In contrast, for the vertically aligned sample, in the first minutes under fog conditions, less quantity of water than in case of tilted sample is collected. Regardless of this difference, collection rate of a vertical sample exceeds the tilted one after 40 min, which indicates that the mass of collected water has increased. Moreover, it can be seen that the amount of collected water increases linearly and constantly during all the experiment.

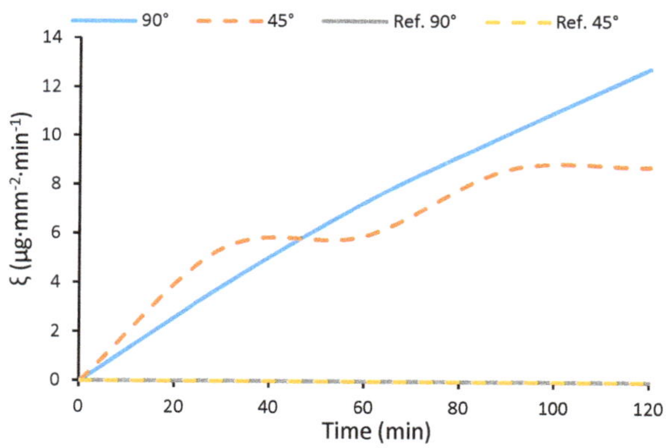

Figure 14. The water collection rate increase in both sample positions, vertically (90°) and tilted (45°) (solid and dotted line, respectively), but it is more pronounced at 90°. References of aluminum substrates (Ref 90° and Ref 45°) are not capable of water collection and are overlapped with x-axis.

Figure 15 depicts coated samples at the beginning of experiment and water droplet condensation after 30 min and 45 min of fog exposure and at the two different tilt angles (90° images at the top row and 45° images at the bottom raw). At 30 min of exposure, water droplets tend to adhere onto the superhydrophobic surfaces and reach an average size of 160 ± 40 µm for both tilt samples. At 45 min of water-harvesting, the size of water droplets at the sample tilted vertically is 286 ± 60 µm and that for 45° is 300 ± 80 µm. Water droplets of the vertically aligned sample (90°) coalesce first, then they increase their size until their diameter (286 ± 60 µm) is exceeded and when they achieve the critical diameter (3.0 ± 0.5 mm), they roll down the surface. Consequently, water droplets leave the surface by gravity and they do it continuously.

As seen in Figure 15, water droplets are adhered to coated surfaces in the initial steps of water collection. It is important to take into account that under fog conditions, water droplet velocity that is increased by the pressure of airstream of fog producer causes them to flatten due to the impact presenting a morphology that can be described as a spherical cap [58]. In fact, water droplets are adhered in planar surfaces until they reach a size of 2.7 mm (according to our result of 3.0 ± 0.5 mm); at this point, by coalescence, the droplets roll off due to the gravity force along the surface [15,59]. In contrast, as the water droplets increase their size, they tend to evaporate, and the water-harvesting efficiency of surfaces significantly decreases. In this experiment, it was found that water droplet diameter at sample tilt 90° is 89% less than that of the maximum size previously mentioned. Thus, this lower size prevents evaporation of water droplets formed onto the coated surface during the water-harvesting process.

Figure 15. Condensation of water droplets on to superhydrophobic coating. The top images correspond to a tilt of 90° while the bottom images are from the sample tilt 45°. All the images have the same magnification.

Having shown the capability of water-harvesting from fog in the case of superhydrophobic surface tilt 90°, its behavior is discussed considering the surface morphology as well as the hierarchical structure of the as-prepared surface. As described previously, the surface is formed by vertically aligned microarrays with a non-linear cone shape that are combined with flower-like structures leading to a hierarchical structure through the whole surface. Water droplet dynamics is governed by Laplace pressure [60]. In a non-linear cone shape, average Laplace pressure will decrease inside the water droplet, which leads to increase its pressure gradient from the tip to the base of the non-linear cone and the water droplet will move away from the tip on a conical surface [61–63]. Then, the gravitational effect increases and improves water collection from fog.

Apart from the key role of microarrays, it is necessary to consider water-harvesting from the hierarchical structure point of view. On one hand, the flower-like structure, i.e., the micro-scale contribution to the system, presents a diameter of 17.5 ± 3.4 µm. On the other hand, the closed packed nanoarrays from the flower-like structure have a thickness of 32.8 ± 2.9 nm. Actually, flower-like structure morphology behaves as a bump feature that maximizes the vapor-phase diffusion on the apex and water droplet condensation increases onto the surface [15,64]. Simultaneously, the close packed nanoarrays of the structure enable the rapid roll-off of the condensed water droplets from the surface [15].

At this point, when water droplets from fog impact the superhydrophobic surface during the first stages they are adhered since the impact of fog fill the air pockets of the surface. Hence, the Wenzel state responsible for water droplet adhesion is created. Despite that fact, as the impact of fog is continuous, water droplets continuously increase the volume enhanced by coalescence of neighboring droplets until a critical diameter is exceeded. Here, as the cap of water droplets is continuously growing because of fog, their morphology changes to a quasi-spherical morphology. This is caused by low adhesive energy of superhydrophobic surface combined with high surface roughness that results in water droplets being easily detached and rolled off. Thus, this behavior leads to a different wetting state assigned to the Cassie regime. This transition from Wenzel state to Cassie state is governed by roughness and the energy of the system related to the position of liquid–vapor interface [65]. Indeed, the more roughness the surface and higher contact angle, the more possible is this transition between both stable states, and because of higher roughness, Cassie state is more favorable in energetic terms. Additionally, the relationship between transition energy and the height of the liquid–vapor interface depends on the nanostructure and the micro-scale features where the latter gives rise to Wenzel–Cassie transition and the former allows the opposite transition. Here, it is shown that water droplets easily detach from the superhydrophobic surface and cause the Wenzel–Cassie transition, which indicates that micro-scale features define this phenomenon, leading to an attractive Cassie state that promotes water repellency and improves water droplet detachment and consequently their roll-off.

4. Conclusions

Herein, we present an innovative sustainable surface prepared onto UNS A91070 aluminum substrate showing WCA = 155°, SA = 5° and a CAH = 1°, which makes it suitable as superhydrophobic surface. Based on magnesium laurate ($Mg(C_{11}H_{20}COO)_2$) this compound contributes to decreasing the surface tension of the surface and together with its lamellar structure, traps the air in the pockets through the whole surface. Moreover, the as-prepared surface is highly robust under SiC abrasive paper test where WCA is kept higher than 150° after every abrasive cycle because of its bulk homogeneity composition, showing self-cleaning properties, as well. In addition, its resistance against UV light is shown, retaining its superhydrophobic properties after 300 min of exposure. The combination of the vertically aligned microarrays in a non-linear cone shape and the intrinsic hierarchical structure of the superhydrophobic surface makes it a very convenient system for water collection in a vertical sample position. This innovative superhydrophobic surface opens a new scope with great potential for water-harvesting based on sustainable reactants that provide a focus on solving worldwide problems of human water consumption.

Author Contributions: Conceptualization, O.R.-A. and N.L.-I.; O.R.-A.; investigation, S.F.-P. and O.R.-A.; data curation, S.F.-P. and O.R.-A.; writing—original draft preparation, O.R.-A., S.F.-P. and N.L.-I.; writing—review and editing, O.R.-A. and N.L.-I.; supervision, O.R.-A. and N.L.-I.; All authors have read and agreed to the published version of the manuscript.

Funding: This research received no external funding.

Acknowledgments: The authors would like to acknowledge the support given by Centres Científics i Tecnològics de la Universitat de Barcelona (CCiTUB).

Conflicts of Interest: The authors declare no conflict of interest.

References

1. Bhushan, B.; Jung, Y.C. Natural and biomimetic artificial surfaces for superhydrophobicity, self-cleaning, low adhesion, and drag reduction. *Prog. Mater. Sci.* **2011**, *56*, 1–108. [CrossRef]
2. Gao, L.; McCarthy, T.J. Contact Angle Hysteresis Explained. *Langmuir* **2006**, *22*, 6234–6237. [CrossRef] [PubMed]
3. Nosonovsky, M.; Bhushan, B. Superhydrophobic surfaces and emerging applications: Non-adhesion, energy, green engineering. *Curr. Opin. Colloid Interface Sci.* **2009**, *14*, 270–280. [CrossRef]
4. Liu, M.; Wang, S.; Jiang, L. Nature-inspired superwettability systems. *Nat. Rev. Mater.* **2017**, *2*, 17036. [CrossRef]
5. Su, B.; Tian, Y.; Jiang, L. Bioinspired Interfaces with Superwettability: From Materials to Chemistry. *J. Am. Chem. Soc.* **2016**, *138*, 1727–1748. [CrossRef]
6. Guix, M.; Orozco, J.; Garcia, M.; Gao, W.; Sattayasamitsathit, S.; Merkoçi, A.; Escarpa, A.; Wang, J. Superhydrophobic Alkanethiol-Coated Microsubmarines for Effective Removal of Oil. *ACS Nano* **2012**, *6*, 4445–4451. [CrossRef] [PubMed]
7. Das, A.; Parbat, D.; Shome, A.; Manna, U. Sustainable Biomimicked Oil/Water Wettability That Performs Under Severe Challenges. *ACS Sustain. Chem. Eng.* **2019**, *7*, 11350–11359. [CrossRef]
8. Mobarakeh, L.F.; Jafari, R.; Farzaneh, M. Robust icephobic, and anticorrosive plasma polymer coating. *Cold Reg. Sci. Technol.* **2018**, *151*, 89–93. [CrossRef]
9. Vazirinasab, E.; Jafari, R.; Momen, G. Application of superhydrophobic coatings as a corrosion barrier: A review. *Surf. Coat. Technol.* **2018**, *341*, 40–56. [CrossRef]
10. Zhuo, Y.; Li, T.; Wang, F.; Håkonsen, V.; Xiao, S.; He, J.; Zhang, Z. An ultra-durable icephobic coating by a molecular pulley. *Soft Matter* **2019**, *15*, 3607–3611. [CrossRef]

11. Brassard, J.-D ; Laforte, J.-L.; Blackburn, C.; Perron, J.; Sarkar, D.K. Silicone based superhydrophobic coating efficient to reduce ice adhesion and accumulation on aluminum under offshore arctic conditions. *Ocean Eng* **2017**, *144*, 135–141. [CrossRef]
12. Masood, M.T.; Zahid, M.; Goldoni, L.; Ceseracciu, L.; Athanassiou, A.; Bayer, I.S. Highly Transparent Polyethylcyanoacrylates from Approved Eco-Friendly Fragrance Materials Demonstrating Excellent Fog-Harvesting and Anti-Wear Properties. *ACS Appl. Mater. Interfaces* **2018**, *10*, 34573–34584. [CrossRef] [PubMed]
13. Dai, X.; Sun, N.; Nielsen, S.O.; Stogin, B.B.; Wang, J.; Yang, S.; Wong, T.-S. Hydrophilic directional slippery rough surfaces for water harvesting. *Sci. Adv.* **2018**, *4*, eaaq0919. [CrossRef]
14. Suvindran, N ; Li, F.; Pan, Y.; Zhao, X. Characterization and Bioreplication of Tradescantia pallida Inspired Biomimetic Superwettability for Dual Way Patterned Water Harvesting. *Adv. Mater. Interfaces* **2018**, *5*, 1–10 [CrossRef]
15. Raut, H.; Ranganath, A.; Baji, A.; Wood, K.L. Bio-inspired hierarchical topography for texture driven fog harvesting. *Appl. Surf. Sci.* **2019**, *465*, 362–368. [CrossRef]
16. Anastas, P.T.; Werner, J. *Green Chemistry: Theory and Practice*; OUP: Oxford, UK, 1998.
17. Erythropel, H.C.C.; Zimmerman, J.B.; De Winter, T.M.; Petitjean, L.; Melnikov, F.; Lam, C.H.; Lounsbury, A.W.; Mellor, K.; Janković, N.Z.; Tu, Q.; et al. The Green ChemisTREE: 20 years after taking root with the 12 principles. *Green Chem.* **2018**, *20*, 1929–1961. [CrossRef]
18. Clark, J.H.; Farmer, T.; Herrero-Davila, L.; Sherwood, J. Circular economy design considerations for research and process development in the chemical sciences. *Green Chem.* **2016**, *18*, 3914–3934. [CrossRef]
19. Klemm, D.; Heublein, B.; Bohn, A.; Fink, H.-P. Cellulose: Fascinating Biopolymer and Sustainable Raw Material. *Angew. Chem. Int. Ed.* **2005**, *44*, 3358–3393. [CrossRef]
20. Calvo-Flores, F.G.; Monteagudo-Arrebola, M.J.; Dobado, J.A.; Isac-García, J. Green and Bio-Based Solvents. *Top. Curr. Chem.* **2018**, *376*, 18. [CrossRef]
21. Spiridon, I.; Darie-Nita, R.N.; Bele, A. New opportunities to valorize biomass wastes into green materials. II. Behaviour to accelerated weathering. *J. Clean. Prod.* **2018**, *172*, 2567–2575. [CrossRef]
22. Zanoletti, A.; Bilo, F.; Depero, L.; Zappa, D.; Bontempi, E. The first sustainable material designed for air particulate matter capture: An introduction to Azure Chemistry. *J. Environ. Manag.* **2018**, *218*, 355–362. [CrossRef] [PubMed]
23. Dichiarante, V.; Milani, R.; Metrangolo, P. Natural surfactants towards a more sustainable fluorine chemistry. *Green Chem.* **2018**, *20*, 13–27. [CrossRef]
24. Schlaich, C.; Yu, L.; Haag, R.; Camacho, L.C.; Wei, Q. Fluorine-free superwetting systems: Construction of environmentally friendly superhydrophilic, superhydrophobic, and slippery surfaces on various substrates. *Polym. Chem.* **2016**, *7*, 7446–7454. [CrossRef]
25. Escobar, A.M.; Llorca-Isern, N. Superhydrophobic coating deposited directly on aluminum. *Appl. Surf. Sci.* **2014**, *305*, 774–782. [CrossRef]
26. Moosavi, S.S.; Norouzbeigi, R.; Velayi, E. Fabrication of flower-like micro/nano dual scale structured copper oxide surfaces: Optimization of self-cleaning properties via Taguchi design. *Appl. Surf. Sci.* **2017**, *422*, 787–797. [CrossRef]
27. Razavi, S.M.R.; Oh, J.; Sett, S.; Feng, L.; Yan, X.; Hoque, M.J.; Liu, A.; Haasch, R.; Masoomi, M.; Bagheri, R.; et al. Superhydrophobic Surfaces Made from Naturally Derived Hydrophobic Materials. *ACS Sustain. Chem. Eng.* **2017**, *5*, 11362–11370. [CrossRef]
28. Selim, M.S.; Elmarakbi, A.; Azzam, A.M.; Shenashen, M.A.; El-Saeed, A.M.; El-Safty, S.A. Eco-friendly design of superhydrophobic nano-magnetite/silicone composites for marine foul-release paints. *Prog. Org. Coat.* **2018**, *116*, 21–34. [CrossRef]
29. Davis, A.; Surdo, S.; Caputo, G.; Bayer, I.S.; Athanassiou, A. Environmentally Benign Production of Stretchable and Robust Superhydrophobic Silicone Monoliths. *ACS Appl. Mater. Interfaces* **2018**, *10*, 2907–2917. [CrossRef]
30. Yue, X.; Zhang, T.; Yang, D.; Qiu, F.; Li, Z. Hybrid aerogels derived from banana peel and waste paper for efficient oil absorption and emulsion separation. *J. Clean. Prod.* **2018**, *199*, 411–419. [CrossRef]
31. Moreno, E.; Cordobilla, R.; Calvet, T.; Cuevas-Diarte, M.A.; Gbabode, G.; Negrier, P.; Mondieig, D.; Oonk, H.A.J. Polymorphism of even saturated carboxylic acids from n-decanoic to n-eicosanoic acid. *New J. Chem.* **2007**, *31*, 947. [CrossRef]

32. Nelson, P.N.; Taylor, R.A. Powder X-ray diffraction, infrared and 13C NMR spectroscopic studies of the homologous series of some solid-state zinc(II) and sodium(I) n-alkanoates. *Spectrochim. Acta Part A: Mol. Biomol. Spectrosc.* **2015**, *138*, 800–806. [CrossRef]
33. Hartman, P. Structural morphology of organic compounds having two centrosymmetric molecules in a monoclinic unit cell. *J. Cryst. Growth* **1991**, *110*, 559–570. [CrossRef]
34. Bond, A.D. On the crystal structures and melting point alternation of the n-alkyl carboxylic acids. *New J. Chem.* **2004**, *28*, 104–114. [CrossRef]
35. Cassie, A.B.D.; Baxter, S. Wettability of porous surfaces. *Trans. Faraday Soc.* **1944**, *40*, 546. [CrossRef]
36. Gadermann, M.; Preston, T.C.; Troster, C.; Signorell, R. Characterization of palmitic and lauric acid aerosols from rapid expansion of supercritical CO_2 solutions. *Mol. Phys.* **2008**, *106*, 945–953. [CrossRef]
37. Caroline, M.L.; Vasudevan, S. Growth and characterization of an organic nonlinear optical material: L-alanine alaninium nitrate. *Mater. Lett.* **2008**, *62*, 2245–2248. [CrossRef]
38. Jiesheng, L.; Yuanyuan, Y.; Xiang, H. Research on the preparation and properties of lauric acid/expanded perlite phase change materials. *Energy Build.* **2016**, *110*, 108–111. [CrossRef]
39. Yin, W.Z.; Tan, Q.; Liu, L.; Li, X.L. Synthesis and Characterization of Mg-Al Layered Double Hydroxide. *Adv. Mater. Res.* **2012**, *454*, 101–104. [CrossRef]
40. Dang, L.; Nai, X.; Zhu, D.; Jing, Y.; Liu, X.; Dong, Y.; Li, W. Study on the mechanism of surface modification of magnesium oxysulfate whisker. *Appl. Surf. Sci.* **2014**, *317*, 325–331. [CrossRef]
41. Robinet, L.; Corbeil-A2, M.-C. The Characterization of Metal Soaps. *Stud. Conserv.* **2003**, *48*, 23–40. [CrossRef]
42. Otero, V.; Sanches, D.; Montagner, C.; Vilarigues, M.; Carlyle, L.; Lopes, J.; Melo, M.J. Characterisation of metal carboxylates by Raman and infrared spectroscopy in works of art. *J. Raman Spectrosc.* **2014**, *45*, 1197–1206. [CrossRef]
43. Saggu, M.; Liu, J.; Patel, A.R. Identification of Subvisible Particles in Biopharmaceutical Formulations Using Raman Spectroscopy Provides Insight into Polysorbate 20 Degradation Pathway. *Pharm. Res.* **2015**, *32*, 2877–2888. [CrossRef]
44. De Gelder, J.; De Gussem, K.; Vandenabeele, P.; Moens, L. Reference database of Raman spectra of biological molecules. *J. Raman Spectrosc.* **2007**, *38*, 1133–1147. [CrossRef]
45. Czamara, K.; Majzner, K.; Pacia, M.; Kochan, K.; Kaczor, A.; Baranska, M. Raman spectroscopy of lipids: A review. *J. Raman Spectrosc.* **2014**, *46*, 4–20. [CrossRef]
46. De Veij, M.; Vandenabeele, P.; De Beer, T.; Remon, J.P.; Moens, L. Reference database of Raman spectra of pharmaceutical excipients. *J. Raman Spectrosc.* **2009**, *40*, 297–307. [CrossRef]
47. Men, S.; Jiang, X.; Xiang, X.; Sun, G.; Yan, Y.; Lyu, Z.; Jin, Y. Synthesis of Cellulose Long-Chain Esters in 1-Butyl-3-methylimidazolium Acetate: Structure-Property Relations. *Polym. Sci. Ser. B* **2018**, *60*, 349–353. [CrossRef]
48. Cocco, F.; Elsener, B.; Fantauzzi, M.; Atzei, D.; Rossi, A. Nanosized surface films on brass alloys by XPS and XAES. *RSC Adv.* **2016**, *6*, 31277–31289. [CrossRef]
49. Wu, P.-Y.; Jiang, Y.-P.; Zhang, Q.-Y.; Jia, Y.; Peng, D.-Y.; Xu, W. Comparative study on arsenate removal mechanism of MgO and MgO/TiO2 composites: FTIR and XPS analysis. *New J. Chem.* **2016**, *40*, 2878–2885. [CrossRef]
50. Dobrovolsky, V.; Khyzhun, O.; Sinelnichenko, A.; Ershova, O.; Solonin, Y. XPS study of influence of exposure to air on thermal stability and kinetics of hydrogen decomposition of MgH 2 films obtained by direct hydrogenation from gaseous phase of metallic Mg. *J. Electron Spectrosc. Relat. Phenom.* **2017**, *215*, 28–35. [CrossRef]
51. Zhang, B.; Li, Y.; Hou, B. One-step electrodeposition fabrication of a superhydrophobic surface on an aluminum substrate with enhanced self-cleaning and anticorrosion properties. *RSC Adv.* **2015**, *5*, 100000–100010. [CrossRef]
52. Chen, L.; Meng, H.; Jiang, L.; Wang, S. Fatty-Acid-Metal-Ion Complexes as Multicolor Superhydrophobic Coating Materials. *Chem.–Asian J.* **2011**, *6*, 1757–1760. [CrossRef] [PubMed]
53. Wang, Z.; Elimelech, M.; Lin, S. Environmental Applications of Interfacial Materials with Special Wettability. *Environ. Sci. Technol.* **2016**, *50*, 2132–2150. [CrossRef] [PubMed]
54. Xiong, J.; Sarkar, D.K.; Chen, X.G. Ultraviolet-Durable Superhydrophobic Nanocomposite Thin Films Based on Cobalt Stearate-Coated TiO2 Nanoparticles Combined with Polymethylhydrosiloxane. *ACS Omega* **2017**, *2*, 8198–8204. [CrossRef]

55. Verho, T.; Bower, C.; Andrew, P.; Franssila, S.; Ikkala, O.; Ras, R.H.A. Mechanically Durable Superhydrophobic Surfaces. *Adv. Mater.* **2010**, *23*, 673–678. [CrossRef]
56. Milionis, A.; Loth, E.; Bayer, I.S. Recent advances in the mechanical durability of superhydrophobic materials. *Adv. Colloid Interface Sci.* **2016**, *229*, 57–79. [CrossRef] [PubMed]
57. Mortazavi, V.; Khonsari, M. On the degradation of superhydrophobic surfaces: A review. *Wear* **2017**, *372*, 145–157. [CrossRef]
58. Welsh, A.; Hou, M.; Meriki, N.; Stevenson, G. Use of Four-Dimensional Analysis of Power Doppler Perfusion Indices to Demonstrate Cardiac Cycle Pulsatility in Fetoplacental Flow. *Ultrasound Med. Boil.* **2012**, *38*, 1345–1351. [CrossRef]
59. De Gennes, P.-G.; Brochard-Wyart, F.; Quéré, D. *Capillarity and Wetting Phenomena*, 1st ed.; Springer: Berlin/Heidelberg, Germany, 2004.
60. Guan, Y.; Cheng, F.; Pan, Z. Superwetting polymeric three dimensional (3D) porous materials for Oil/Water separation: A Review. *Coatings* **2019**, *11*, 806. [CrossRef]
61. Lin, J.; Tan, X.; Shi, T.; Tang, Z.; Liao, G. Leaf Vein-Inspired Hierarchical Wedge-Shaped Tracks on Flexible Substrate for Enhanced Directional Water Collection. *ACS Appl. Mater. Interfaces* **2018**, *10*, 44815–44824. [CrossRef]
62. Zhang, S.; Huang, J.; Chen, Z.; Lai, Y. Bioinspired Special Wettability Surfaces: From Fundamental Research to Water Harvesting Applications. *Small* **2016**, *13*, 1602992. [CrossRef]
63. Gurera, D.; Bhushan, B. Bioinspired conical design for efficient water collection from fog. *Philos. Trans. R. Soc. A Math. Phys. Eng. Sci.* **2019**, *377*, 20190125. [CrossRef] [PubMed]
64. Park, K.-C.; Kim, P.; Grinthal, A.; He, N.; Fox, D.; Weaver, J.C.; Aizenberg, J. Condensation on slippery asymmetric bumps. *Nature* **2016**, *531*, 78–82. [CrossRef] [PubMed]
65. Nosonovsky, M.; Bhushan, B. Patterned Nonadhesive Surfaces: Superhydrophobicity and Wetting Regime Transitions. *Langmuir* **2008**, *24*, 1525–1533. [CrossRef] [PubMed]

MDPI
St. Alban-Anlage 66
4052 Basel
Switzerland
Tel. +41 61 683 77 34
Fax +41 61 302 89 18
www.mdpi.com

Coatings Editorial Office
E-mail: coatings@mdpi.com
www.mdpi.com/journal/coatings

www.ingramcontent.com/pod-product-compliance
Lightning Source LLC
Chambersburg PA
CBHW062107220926
51850CB00010B/2920